D1499778

INTERPRETATION
IN SONG

Da Capo Press Music Reprint Series

MUSIC EDITOR
BEA FRIEDLAND
Ph.D., City University of New York

INTERPRETATION
IN SONG

BY

HARRY PLUNKET GREENE

With a New Introduction by
DOROTHY URIS

DA CAPO PRESS • NEW YORK • 1979

Library of Congress Cataloging in Publication Data

Greene, Harry Plunket, 1865-1936.
    Interpretation in song.

    (Da Capo Press music reprint series)
    Reprint of the 1912 ed. published by Macmillan Co.,
New York.
    Includes index.
    1. Singing—Interpretation (Phrasing, dynamics, etc.)
I. Title.
MT892.G74   1979   784.9'34   79-4135
ISBN 0-306-79509-4

Published by Da Capo Press, Inc.
A Subsidiary of Plenum Publishing Corporation
227 West 17th Street, New York, N.Y. 10011

INTRODUCTION TO THE
DA CAPO EDITION

Greene, Harry Plunket (*b.* Old Connaught House, Co. Wicklow, 24 June 1865; *d.* London, 19 Aug. 1936) . . .

"His remarkable powers of interpretation and especially, the beauty of his enunciation in singing the English language made him one of the leading apostles of English song from folk-song to the products of modern composers . . . His book, *'Interpretation in Song'*, contains the essence of his teachings."
—*Grove's Dictionary of*
Music and Musicians
Fifth Edition

I welcome the opportunity in preparing these introductory comments to reread and re-relish *Interpretation in Song*. In the process of research for my own book, *To Sing in English,* I had found the paucity and content of material on the subject quite disheartening until Harry Plunket Greene came to the rescue. His book was for me a beacon of clarity in a murky scene. In the year 1912, he produced this forthright work on the ways of song. Into it, he poured a teacher's fervor and an Irishman's faith in the mother tongue as a supremely singable language. We chuckle as we read his witty gibes at the music world of his day.

The book's reissue is well-timed. Performance in English of opera, art songs, and oratorio, old and new, is expanding at a rate inconceivable ten years ago. Performers have greater purpose than ever to project the language intelligibly and artistically. And what with opera and other works now on the home screen—English words had better come clear!

As artist and teacher, Harry Plunket Greene's guidance in the interpretation of vocal music remains ever fresh. Reading his book today revives fundamental and enduring concepts for singers and their mentors. "Every song has an atmosphere of its own—something all-pervading to which detail is subordinate and at the same time to which every detail contributes . . . It's not what can be put into the song but what can be got out of it." Shades of Stanislavski!

Among other guidelines, his *Three Rules* instruct us to "never stop the march of a song," and to "sing mentally through your rests," and to "sing as you speak." In the fustian period in which he lived, PG (to his friends) had the taste and courage to decry the rampant use of cheap effects, the elaboration of detail, and the performer's overweening concern with self rather than art and audience.

Lest this discussion begin to sound like an unalloyed hymn of praise, let me point out that PG's conventional "making of programmes" for recitals would not pass muster with critics today.

Neither would his somewhat simplistic ideas on vocal technique meet present standards of voice training—such as, for example, the advice to lift the chest as high as it will go and keep it there. Though in many respects a radical in his period, he may have sought to soothe the traditionalists by conservative views on music. In any case, the differences we may have with him are minor whereas his interpretive and linguistic techniques continue to have major validity in our time.

While PG's skill with other languages was notable, especially with German (the book treats lieder in depth), it was the purity of his sung English that created the greatest stir. Almost a hundred years ago, he performed to acclaim his first English role in Handel's *Messiah*. Interpretation as an art inevitably dovetails with diction because the two, asserts the author, are inseparable.

When PG advocates *sing as you speak,* he makes clear that he does not mean the "mumbled jargon of British conversation," any more than we mean the slipshod variety often heard in the United States. He goes on to say that ordinary talk is not "speaking in its true sense . . . the text must be as clear-cut and clean as the music it adorns. All the singer's perfection of techniques goes for little or nothing if his singing is not speech in song." Because of our native twangs and mutterings, we have had to defend the singability of English against its critics who base their prejudice on local speech patterns. What nonsense! Alive and well, American English long ago came of age as a spoken, written, and *sung* language. And *good*

American speech can breathe new life into the whole spectrum of vocal literature.

PG tells us not to blame singers for poor diction, rather to blame the public who makes no demands on them. As to that, I hear more complaints about incoherence from those out front, and from the critics' reviews which now tend to take English-as-sung into serious account; performers have come to rely less on mere vocal display and more on the text's meaning, and recital programs do not invariably, as in the past, place a few token English songs at the end of a program. Indeed, I have attended lately several recitals where English actually not only opens the program but may also include a song cycle in our native tongue.

PG faults the "cult of the open vowel" for much of the mischief done the English vowel. All vowels, like satellites, circle around *ah* so that *man* is sung like *mahn*, a misbegotten vowel neither English nor Italian. The author exhorts singers to avoid "the hybrid vowel and the shirked consonant." When delivered in a mishmash of accents and distortions, such diction resembles no known dialect heard anywhere on earth except on concert and opera stages.

We know that ours is more strongly accented than most other languages of the vocal repertoire. Its special genius lies in vivid contrasts of strong and weak word-types and syllables, and stress and unstress are vital for transmitting signals of meaning to the listener. PG, to his credit, holds fast to this basic linguistic principle as the foundation of verbal communication in song. How he applied the structure of the language so brilliantly to singing in English places him far above his contemporaries. "If singing is to be speech in song, it must primarily talk sense . . . To talk sense intelligibly, the talker, be he speaker or singer, must give his words their right values according to the rules of prosody."

When musical patterns obscure English stress patterns, we teachers realize that singers have verbal hurdles to surmount. Seduced by the score, however, too many of them flatten the obligatory contrasts within the text—an action fatal to meaning. Let them remember that, when performers fail as interpreters, they do so largely through the neglect of the word rhythms of English; instead, their skillful use adds another dimension to song. "Why should a word set to two equal notes be sung with even value?" asks the author, "equal time value, yes—but not necessarily equal *pressure-values*" (an excellent term). Not *glor-ee,* but *glo*ry; not *bee-fall,* but be*fall;* not *beauteef*ul, but *beaut*iful. "As often as not, the note can be brought into line and given its right word value with no loss to the power of the phrase . . . To the master of style, the dovetailing of the two—phrasing and sense—is a labor of love."

What PG calls the "identity of texture" is also required of the sung and spoken word. In short, a recognizable word in song cannot deviate from its spoken form, the indigenous sounds must match since the language is the same for singer and hearer. How exhilarating is an audience's response to this shared experience of word identity!

And in turn, the artist, aglow with appreciation, reaches out to them. Should a singer have difficulty adjusting the text to the score, PG advises intoning the words on a comfortable pitch without worry about phrasing and to keep at it, word by word, until he can transplant them, texture and all, into the song.

The author's treatment of the consonant, that bogie of singers, is altogether admirable. "We are not to tolerate our consonant or make the best of him. We are to use him as a friend." We are still waging the battle for our friend, the consonant, and PG gives us plenty of ammunition. First of all, he makes much of "word-illustration," his term for the expressive singing of the consonant as a prime aid to interpretation. Then he extols the color and dynamic impact of words like *c*urse, *sh*ame, *dr*eam, blu*sh*, dro*ne* ("Note the *n* in *drone* is far more expressive than any holding of the vowel.") Actively pictorial, these vital shapes perform lavishly as onomatopoeia in our tongue. What a boon to interpreters are such words as: *hiss, tinkle, bang, whisper,* and so on and on!

In reconciling the phrasing to the text, a singer has varied means (apart from clear articulation) to clarify and enhance diction: tone-color, word-illustration, dynamics, pauses, rubato, and other interpretive effects. The directive to "sing mentally through the rests" also supports the overall sense of the text. The *Kunst-pause* or artistic pause is a quick stop before a word in need of emphasis, the same device employed in speaking poetry. To the artist who seeks to communicate above all, the creative touches in song have always meant an everlasting search.

The final chapter, *The Straight Line in Phrasing,* was delivered as a lecture in 1917, and sums up the book in PG's jauntiest style. He borrows enthusiastically from the French their distinctive technique of legato singing—the forward drive of rhythm with the consonant in the lead. Thus, he points out, the consonant belongs *not* to the vowel it comes after but to the vowel that follows it. "Must consonants interrupt the forward movement of the vowel? God forbid!—they must enhance it." As the consonants supply the push, the melody is carried on the vowels and the song moves onward with "little hammers and every consonant a hammer" along "a railway line irresistibly."

All this is to the good. We may indeed learn from the French how to bring words together so that the listener comprehends with ease what we are attempting to convey. For English to be intelligible and beautiful, consonants must also join vowels in a non-stop flow of sound and sense. However, our accent patterns are quite different from those of the French: for example, "je vous aime" spoken or sung, becomes "je vou zaime," the final *z* sound in *vous,* carried over, becomes the strong initial consonant in *aime.* When we link a final consonant in English to the next word or syllable (and of course, we must), we carry it over no more strongly than it is pronounced in the given word. Therefore, "those eyes" is never "tho *z*eyes" but rather "tho-*z*eyes." We retain the stress on the initial vowel in the noun *eyes* and carry over *z* unstressed.

Yes, English is a legato language and the linkage of its sounds calls for dexterity. While we applaud PG's fervent advocacy of the "straight line in phrasing," let us always remember our underlying word rhythms as we send the consonant ever ahead, and the next one and the next, until the song is done.

How PG inveighs against "the fetish of freedom of the jaw" which has been made "the bond-slave of the would-be singer!" "The chewing of words" has led "to the degradation of tone and the utter destruction of rhythm." Instead of the jaw, singers had best devote themselves to "mobility of the lips." I find his prescription for proficient diction remarkably up-to-date. "Pronunciation is done practically entirely by fine, closely associated movements of the lips and tip of the tongue around the teeth." And in this urgent admonition, I hear the echo of myself: "If diction takes place further back in the mouth, it is wrong."

What our farseeing author does not cover in this book, diction teachers must add: the essential breakdown of the strong and weak word-types of English in a specific system of study for English texts; the *production* of the consonants he values so highly, and an awareness of the important differences among the twenty-five shapings of sounds. There are other omissions and emphases to which teachers and coaches may take exception. But *he* has given us this book—a groundwork, well-cultivated, for future growth. We have caught fire from his impassioned devotion to communication in song, its text and its interpretation, and, as he says, we will not forget that "music's the thing."

May I indulge myself, in conclusion, with a favorite quotation from a famed contemporary of Harry Plunket Greene:

> If you wish to sing beautifully—and you all do—you must love music; and the nearer you get to music the more you will love it. If you wish to sing your native language beautifully—and you should—you must love your native language; and the nearer you get to it the more you will love it . . .
>
> —*Nellie Melba,* from the text of a lecture delivered at the Guild School of Music, London, about 1913.

DOROTHY URIS
New York City
December, 1978

INTERPRETATION IN SONG

THE MACMILLAN COMPANY
NEW YORK · BOSTON · CHICAGO
DALLAS · SAN FRANCISCO

MACMILLAN & CO., Limited
LONDON · BOMBAY · CALCUTTA
MELBOURNE

THE MACMILLAN CO. OF CANADA, Ltd.
TORONTO

INTERPRETATION IN SONG

BY

HARRY PLUNKET GREENE

Das muss ein schlechter Mül-ler sein, dem
nie - mals fiel das Wan-dern ein.

New York

THE MACMILLAN COMPANY

1912

To

FRANCIS KORBAY

AND

SAMUEL LIDDLE

CONTENTS

PAGE

INTRODUCTION ix

PART I

EQUIPMENT. 1
    Technique — Magnetism — Atmosphere — Tone-
    colour — Style.

PART II

RULES :
    Main Rule I. 37
    Main Rule II. 92
    Main Rule III. 104

PART III

MISCELLANEOUS POINTS 145
    Styles of Technique — The Singing of Recitative —
    Pauses — *Rubato* — Carrying-over — The *Melisma* —
    The Finish of a Song — Consistency — Word-illus-
    tration — Expression-marks — Conclusion.

PART IV

THE CLASSIFICATION OF SONGS — SONG-CYCLES — THE
    SINGING OF FOLK-SONGS 198

PART V

THE MAKING OF PROGRAMMES 223

PART VI

HOW TO STUDY A SONG 233

APPENDIX

HOW TO BREATHE — THE CLERGY AND INTONING . 289

INDEX 301

INTRODUCTION

It is a popular fallacy that a beautiful voice is synonymous with a lucrative profession and entitles its possessor to a place among the masters of music. England is full of such voices, in various stages of technical training; some full of hope for the career ahead, some despondent and puzzled at the nonfulfilment of that hope, and others — a vast number — for whom hope is dead and the grim struggle for a livelihood the only question. To such it seems inconceivable that a thing of intrinsic beauty, a great gift like a voice, should count for nothing in the world of music, and the singer in his disappointment attributes his failure to the shortcomings of his manager, the opportunities of his rivals, the personal prejudices of his critics or the relentlessness of the gods — to anything but the true cause. The explanation is simple enough — he has not learned his business. With the minimum of efficiency he has assumed the maximum of responsibility. While still speaking the language of his childhood, he has ventured out into the world to take his place among men.

He has every excuse so to do. Born in the land of cricket and fair play, by the very privileges and responsibilities of his birthright he has become the best fellow of his class in the world; but insularity has its disadvantages, and the English singer of limited means has no chance to rub shoulders with his colleagues abroad or widen his horizon. The atmosphere of foreign student life has never entered into his soul. He knows no language but his own. For him there is no National or Municipal Opera wherein to hear the masterpieces of music or take his place as interpreter. His outlook is bounded by the conventional oratorio and the "popular" ballad, and between the respectable oases of the one and the miasmatic swamps of the other he wanders through the desert. During no period of our musical history have technique and invention made such strides as in the last generation. The Wagner score of thirty years ago, the terror of the orchestral player and the wonder of the composer, has become the commonplace of the one and the text-book of the other. Each has risen to the level of his responsibilities and played a man's part. The singer alone has stood still. The reasons need not be discussed here. The fact remains that our platforms are overrun with voices half-developed and quarter-trained, singers without technique, without charm and without style, to whom rhythm is of no account and language but the dead vehicle of sound, whose ambitions soar no higher than the three-verse song with organ obbligato, and to whom the high-note at the end and the clapping of hands spell the sublimity of achievement.

The singer with a beautiful voice who has not taken the trouble to learn his business is a commonplace; his prototype is to be found in every profession in the world. His standards are perforce on a level with his proficiency. Sufficient unto the song is the singing thereof, and by his applause he measures his musical stature. But when the song comes to its inglorious end, both song and singer are thrown out together into the world's rubbish-heap.

He has no cause to complain. One glance at the story of music would show him that in the scheme of things he is of no account. It is the composer who lives; the singer is one of the Ephemeridae. Invaluable to anecdote, immaterial to history, he belongs to reminiscence not to record.

Who was the great tenor of the Thomas-Kirche? There is no memorial even of his name; but Bach who wrote the great arias for him is with us for all time. If the tenor of the Church cantatas is buried and forgotten, surely the singer of the British ballad is justified in dying young.

There is another far more dangerous than he; the man who with great gifts, competent training, and full knowledge deliberately uses his powers for degraded ends. Such a man is not only a traitor to his art but a menace to society, for the public looks to him for guidance and follows in his path. The greater the man, the greater the responsibility and the greater the crime. Unlike the other, he trusts that history will let him alone.

There is yet one more — the man who, whatever his gifts, whatever his opportunities, means to play the game. What hard words he sees in these pages are not for him, though for him the book is written. It does not profess to represent more than the personal opinions of its author. Those opinions are the result, arrived at by a process of analysis, of a good many years' hard work in public and harder work at home.

They are the experiences of one who has been through the mill, who by loyalty to his rules has tried hard to atone for the shortcomings of his equipment, and they are primarily meant for the man who having served his apprenticeship is starting out into the world on his own. There are no short cuts in art, but if it helps to make the road to the dim Parnassus any smoother, or gives him a tune to whistle to his stride or, best of all, shows him fresh lands to explore, the book will have served its purpose.

It does not pretend to be an exhaustive treatise on its subject. Its object is to give in the shortest possible form what is most likely to prove useful. For obvious reasons the examples quoted are those with which the author is most intimately acquainted, and for equally obvious reasons (of copyright) those examples have been drawn mostly from the classics. But the treatment of a song is the same whether it lie high or low, be old or new, or sung by man or woman.

Interpretation knows no restrictions of compass, age or sex.

H. P. G.

INTERPRETATION IN SONG

INTERPRETATION IN SONG

PART I

EQUIPMENT

INTERPRETATION is the highest branch of the singer's art. To that end he has worked, and when he has reached it he has begun to live. The wise master, when his pupil's wings have grown, will let him fly; he will cease to dictate and begin to collaborate. Interpretation is the highest branch, after the creative, of every art, but the singer has greater privileges than his fellows, for it is given to him to interpret to his fellow-men the great human emotions in the language of the poets ennobled by music. How to express those emotions in that language in the finest way is the fascinating problem before him. His field is limitless, for the masterpieces of song are inexhaustible and, if he prove himself worthy, the architects of his own time may choose him for their master-builder. But if he has privileges, he has responsibilities. Every time he sings he assumes the guardianship of another man's property. It is committed to his care on trust and on him may depend its fate. Any singer who has taken part in bringing out a new work at a great festival will appreciate the weight of that burden.

The further the singer advances in his art, the higher the place which study takes in comparison with

B   1

performance. The study of Interpretation is mainly intellectual and psychological, its actual performance largely physical and dependent on outside conditions. The hundred and one drawbacks of a concert room may bring the best laid schemes to grief. Public performance, even the most successful, is like the salmon on the bank — a record of achievement with the best of the fun over. The song once learnt and sung is never quite the same fairy-tale of romance; the child has grown up and left his fairy-tales behind him.

What more exciting moment could the singer ask than the first step in the study of a great song-cycle such as Schumann's "Dichterliebe," the assimilation of each song in turn and its moulding to the general scheme; or the absorbing of the atmosphere of Schubert's "Leiermann," the visualising of the picture of the poor old hurdy-gurdy man, barefoot on the ice, grinding out the same old tune, unheeded and unheeding; or the dramatising in song of the father, the child and the Erl King; or the Knight and the Lorelei in Schumann's "Waldesgespräch"; or the suggesting of the remoteness of Stanford's "Fairy Lough"; or the heat-haze and human throb of Vaughan Williams' "Silent Noon"? Every one a little drama, or picture, or colour study, and every one a masterpiece; and for every treasure that he discovers he knows the earth to hold a thousand more.

Songs are the property of a commune of individualists. They belong to all alike, and each is the private possession of the singer who sings it. A poem may have a perfectly different emotional effect on one man and on another. If that is true of words, how much more must it apply to music, and how much more, again, to words and music together. The combination of the two makes, perhaps, the strongest emotional appeal that we know to the individual, and his response thereto depends on his temperament, intelligence and equipment. In no two men are these alike. Interpretation is, therefore, essentially *individual*.

It is well that it is so, for if it became stereotyped there would be no scope for personality; imagination would count for nothing and originality would be a dead letter. Songs would be moulded to a pattern, and the tyranny of the beautiful voice would be established for good, while its possessor would become, if possible, more insufferable than he is already. "Individuality" is the singer's greatest asset. Every song in the world is his property to do what he likes with. So much the greater his responsibility.

But what of the composer? Has he no say in this piratical appropriation? It is the charlatan whom the composer fears, the cheap-jack who juggles with the rhythm and reads in false effects to gain applause. The man of enterprise is his friend.

There is a fascination about the pioneer; he carries romance in his very name. The story of Columbus or the North-West Passage, or even the Odyssey, makes, rightly or wrongly, a more vivid appeal to the imagination than the life of any statesman or the fifteen decisive battles of the world. To explore the unexplored, to "walk out toward the unknown region," to win the secrets of the earth by force of arms have been the dream of boy and man from time immemorial. As with Nature, so with Art. Song has her dark continents and virgin peaks waiting to be conquered, and when they call the pioneer must be up and doing. But let him see to it that he starts prepared. Voice alone, however great, will not carry him far, nor enterprise alone without equipment. That equipment comprises various gifts to which the public, and possibly the singer himself, have never given a name, and requires years of training to fit him to apply them. All are essential to true success.

The interpreter must start with four possessions:

> Perfected Technique.
> Magnetism.
> Sense of Atmosphere.
> Command of Tone-colour.

Of these the first can be acquired by anybody, the second is a pure gift, while as regards the other two the singer can either be born with them or assimilate them successfully by study and imitation.

PERFECTED TECHNIQUE

Technique is easy to acquire; it is difficult to absorb. The one is a matter of months, the other of years. Technique is not uniform, and in its rules does not apply to each individual alike, but any man of intelligence can be taught the one that suits him best. It is the absorption of that knowledge into his very system so thoroughly that its application becomes automatic which is the difficulty. The physical use of his voice must be the *unconscious response to the play of his feeling*. That is a matter of years, and few singers have the patience to see it through.

It is obvious that interpretation would be impossible if the singer were hampered by difficulties of actual performance. If his mind be absorbed by the physical struggle, the intellectual side cannot have fair play. Nay more, the greater his gifts, the greater his danger. Temperament is a bad horse to ride if you cannot control him; sooner or later he will break your neck. Technique must be the singer's servant, not his master, and must follow his mind as automatically as his hand follows his eye. If not, attack is turned into defence. It is a long and wearisome business, for there are no short cuts. The freak has yet to be discovered who grew six feet in the cradle.

Anyone, then, can acquire technique; but even technique has its individuality. To some it seems to present no difficulties — control comes naturally and progresses by easy stages; to others it is a matter of extreme laboriousness. Again, the same style of technique does not suit everybody; the main lines are universal, but the particulars have to be humoured to the individual. This may be due partly to the physical formation of his vocal organs, but partly, no doubt, to his temperamental peculiarities. Their recognition and handling are the work of the teacher. The sounder he makes the early work, the better the workman he turns out. If the pupil's apprenticeship has not been truly served, he will surely come to grief in the end. Sooner or later he will be faced by a wall which he cannot get over; then back to the beginning he must go — with a broken heart — or settle down to mediocrity.

It is a commonplace to say that a public singer must be possessed of physique. In these days of huge *répertoires* and long recital programmes, where the performer is on his feet for the better part of an hour and a half, singing twenty to thirty songs of every shade of emotional effect, the strain on heart, lungs and brain is abnormal, and without physique he could never last out. But physique alone will not pull him through. He may blunder through half a dozen songs by main force, but if his technique is not his servant he will not win half-way to the end.

One word of advice may be given to the beginner. Let him avoid the voice-production faddist and purveyor of short cuts. There are no short cuts. Let him likewise avoid the anatomical-jargon man. Lungs, chest, nose, palate, tongue, teeth and lips are the singer's stock-in-trade and need no diagrams. The less he knows about their physical details the better. Anatomical illustrations and treatises will not place his voice a whit better. They will merely render him self-conscious and worry him into senseless solicitude about organs whose movements are mainly automatic.

For the purposes of this book perfected technique must be assumed. The singer's complete command of his physical powers must be taken for granted; interpretation has to do with the intellectual uses to which these powers are put. For this reason no actual physical directions will be given in regard to any of them, with one exception — *Breathing*. Correct breathing is perfectly simple. It should not present the smallest difficulty to anybody : yet there is no part of the business over which the faddists have run such riot. As the whole structure of interpretative singing stands upon breath-control — its very foundation — the physical part of breathing will be dealt with in a special appendix (p. 289).

In order to interpret, the singer must have at his command:

1. Deep breathing and control of the breath.
2. Forward, and consequently resonant, "production" of the voice.
3. The power to pronounce pure vowels and distinct consonants with ease.
4. The power to move at any pace with ease.
5. The power of phrasing — both long and short — with ease.

In explanation of these it may be said:

(1) Deep breathing has to do with the taking-in of breath, control with the letting of it out, and of these the second is infinitely the more important.

(2) The singer should be unconscious of the fact that he possesses a throat. Where his voice is *produced* goodness knows; the singer certainly does not. Where it sounds, and rings, is his business. If it sounds in his throat, it certainly will not ring.

(3) This is closely associated with No. 2. Power of diction depends upon the use of the tongue, teeth and lips. It is, therefore, of primary importance that the material they work with should be at their disposal, and not fighting its way through a museum of uvulas and epiglottises and other anatomical specimens.

(4) This is closely associated with all three of the foregoing. If these are above proof, singing fast will be as easy as singing slow. It will in fact be easier, as involving slightly less physical control of the breath. It may appear paradoxical, but to a singer with his technique at his disposal singing fast is a rest, so much so that a tendency to sing fast is an

outward and visible sign of laziness and avoidance of difficulties.

(5) Phrases are the bricks out of which interpretation is built, phrasing the laying of these bricks by the master-builder. On the knowledge of his business (Nos. 1, 2, 3, 4 above) depends the laying (No. 5).

The key to them all is contained in the words *with ease.* Technique is the means, interpretation the end. The end will be difficult if the means are not easy.

Finally, the singer should bear in mind that he is not the best judge of his own technique and that that technique can never be left alone. The physical part of his work should from time to time be submitted to other expert inspection. No matter how high he may stand in his art, his technique is bound occasionally to grow rusty or get out of control. In such cases he will do well to submit it to a master whom he can trust. The doctor when he is ill consults a colleague. The singer, where the health of his voice is concerned, should follow his example.

MAGNETISM

Any singer who has sincerity, a fair amount of imagination and perfected technique can interpret, but not necessarily successfully. To be successful he must have Magnetism. Magnetism — so-called for want of a better word — is a pure gift. It is as much born to the individual as the colour of his eyes or his hair. It is the property in greater or less degree of every successful public man, be he preacher or politician, actor or singer. It is probably an applied form of what in the private individual is called "attraction," though its application is unconscious and spontaneous. It is generally closely associated with temperament though in no way related to it, for the one may appear without the other at any time. It is the greatest gift the singer can have, for its possession means power.

Magnetism is the indefinable *something* which passes from singer to audience and audience to singer alike; for the audience which the singer holds in the hollow of his hand, holds him as surely in its own. Each acts and reacts on the other in ever-increasing degree. It is a gossamer thread over which passes that nameless electric current which stirs the singer to his depths and holds his audience thrilled and still. Which starts to spin it we do not know; probably the singer, for one man has it and another has not, and when he has not, no audience can give it to him. The man who has it is unconscious of exerting it. He knows when it is present and when it is absent — it is as fickle as a will-o'-the-wisp — but he can never deliberately start it on its journey. All he knows is that when it is there it seems to grow and as it grows it intoxicates. It is worth all the applause in the world; applause is but its by-product. Its absence leaves an almost physical feeling of depression. It may be apparently absent and suddenly appear. One friendly face or understanding personality in an audience may suddenly set it going; some external or ludicrous incident may bring a smile to every face, and along that smile will run the little thrill which set singer and audience calling one another by their Christian names. It may never be there, and it may be there from the beginning. It likes silence — the wise man will not begin his song too soon. It

likes attention — he will treat it with deference. It
likes concentration — he will see that his thoughts are
not elsewhere. The wise man knows that if its friends
are many, its enemies are legion. To the experienced
singer it is always a mystery how it manages to
survive through a single song. Every man's hand is
against it. A breath will blow it away; the banging
of a door, the dropping of an umbrella, the rustling of
programmes (the wise man will so print his books of
words that there is no turning over in the middle
of a song), all the thousand and one accidents which
distract the attention of audience and singer alike, all
are its deadly enemies. Strange to say, things heard
are far less distracting than things seen. Church
bells, motor-horns, muffin-men, barrel-organs, dogs,
seem to concentrate round concert halls; Magnetism out
of sheer familiarity pulls a long nose at them. A
pistol shot or a braying donkey might through very
explosiveness or appropriateness bring the singer to
confusion, but the ordinary run of noises passes him
and Magnetism by. But let an old gentleman in the
front row stand up and take off his overcoat, or let a
programme boy meander up the middle aisle, or, worst
of all, let a late-comer open the door and walk to his
seat; every eye leaves the platform, snap goes the
magnetic thread, and the singer comes down and breaks
his crown, and the song comes tumbling after.

The late-comer — and in a lesser degree the early-
goer (what the late Hans von Bülow called the
Spätlinge and the *Frühlinge*) — is Magnetism's deadly
adversary. The human being has yet to be found
with a will strong enough to keep the human eye
away from him. Times out of mind the late-comer

has wrecked both singer and song, and for that penance many of the singer's sins may surely be forgiven him both now and hereafter. There are still some survivors of the Stone Age who, because they have bought a ticket for a concert, consider that they are thereby entitled to disturb the audience and ruin the performance at will. It has never occurred to them that the interests of the performer and their own are identical. To them any desire on the part of the singer to play fair to his audience and his work is looked upon as a presumption. Fortunately, however, the public has ceased to see the fun of it, and in self-protection now demands to listen to each number with closed doors. The late-comer smuggles through nowadays by inadvertence.

One fact emerges from all this, that to Magnetism the most important medium for good or evil is the eye. To the singer the roving eye in an audience is as terrible a danger-signal as the early yawn. If all eyes are on him he knows he is all right. If they wander by, taking him casually on their way, he is probably all wrong, and equally probably it is his own fault. If they are fixed upon him and leave him with a jump, the fault is generally the late-comer's. In either case, no matter whose the fault, his work is of no account.

What then can the singer do to help the eye, Magnetism's best friend? He can see that that friend is not employed elsewhere, and be ready to throw the door open should Magnetism look in, after its wont, through the keyhole. The eye that is fixed upon the printed page is no good to it. No amount of make-believe on a London November day could conjure up the blue skies and white clouds of the midsummer of "Feldeinsamkeit" while the singer's head was bobbing up and down from the vocal score. He wants his eyes for something else, not only to visualise his picture, to look out *unconsciously* — for the eye follows the mind as unconsciously as the hand follows the eye — to those green fields and summer skies, but to gather from, and give back to, his audience that indescribable magnetic sympathy, communicable as much by the eye as by the clapping of hands, which makes them both friends, coaxes away the terrors of nervousness and sets the light to his imagination.

The interpreter must memorise his work.

Whatever may apply to instrumentalists and abstract music, this rule is vital to singing. Song deals with the great human emotions expressed in words, and the singer stands face to face with his audience. Every friend of expression that has been given him he is in duty bound to make the most of. Hard work is not easy. Memorising is a work of extreme laboriousness, but when that work is done it is the singer's possession for ever. Nay more, it has a power all its own of separating the sheep from the goats. To memorise a poor song is a martyrdom; the eclectic faculty shrinks from the bad and the shallow. It is a subconscious faculty working independently and it is no respecter of persons. Time and again the singer will find that in the memorising process it has stultified his own earlier judgment and treated a great name as of no account. In the furnace of its refining fire the dross is burnt away and only the pure gold remains.

SENSE OF ATMOSPHERE

Every song has an *Atmosphere* of its own; that is, a something all-pervading to which all detail is subordinate and to which at the same time every detail contributes.

It follows that every song must be treated *as a whole*.

The composer wrote it as a whole; the singer must sing it as a whole. A musical phrase is made up of a number of notes. The singer does not think of those notes separately; he thinks of the phrase as a whole, and the song is to the phrase what the phrase is to the note. The mind absorbs the picture, and the detail fits into perspective of itself.

This treatment of the song *as a whole* is the secret of interpretation.

The treatment of its Atmosphere demands an attitude or Mood. The mood belongs to the singer, the atmosphere to the song, and the best friend of both — the father of the one and the godfather of the other — is Imagination.

Some songs breathe their atmosphere in every bar; in others it is so subtle it cannot be given a name. Imagination will christen it. If the singer has imagination, atmosphere will come to meet him half-way. Its fascination will lure him on and lead him into fairyland. A thousand feet may have worn the path bare before him, but to him all is virgin soil. In after life the memory of that moment will thrill him, when first the mists lifted, and the *Wanderlust* entered into his soul.

Imagination is many-coloured. He must have all sorts; not merely the power to visualise his scene, to paint upon his inner vision the picture of the father, the son and the "Erlking," or the horror of "Gruppe aus dem Tartarus"; the drowsy blue day of "Feldein-samkeit," or the dragon-fly of "Silent Noon" — "hung like a blue thread loosened from the sky" — but a belief in happiness, in Leprechauns and fairies, in Santa Claus and Father Christmas, and a deep love of nature and children.

If his imagination can show him the atmosphere, the mood follows of itself.   In Brahms's "Feldein-samkeit," for instance, the *atmosphere* is one of dreamy happiness, of utter contentment mental and physical; the *mood* one of laziness, of half-closed eyes, of some one hypnotised by the hum of bees and drugged with the scent of flowers.   Given that mood the song sings itself.   It tells, it is true, of long green grass, of the ceaseless hum of insects, of blue skies and white clouds like floating dreams, but the singer does not think of them.   They are details; they simply contribute to the *atmosphere* of the song *as a whole*.   He tells you of his *mood* — "as though he long were dead and borne along to heaven" — happy, lazy, half asleep.   Let him but accentuate the detail or worry over his technique, and the skies will turn to thunderstorms, the bumble-bees to mosquitoes and the white clouds to water-spouts.

It is not necessary to ticket either atmosphere or mood with a name.   The singer need only realise the meaning.   Thus, in another song of Brahms, "Auf dem Kirchhofe," there are two atmospheres — the storms of life, with detailed expression in the rain and wind beating against the old tombstones and withered wreaths and worn inscriptions, followed by the calm of those who lie beneath; the contrast summed up in the words *gewesen, genesen* (an antithesis practically impossible to render into English). It would be hard to define in words either atmosphere or either mood. The singer would simply be conscious of a spirit first of inevitableness and despair, and then of redemption and peace, and fit his moods to each.

In neither of the two songs quoted above is there any drama to act, any incident to relate; only with atmosphere, mood and colour to express in the one the drowsy summer's day, and in the other the poet's epitome of life and death.

Some songs carry their atmosphere and mood in their very name. Schubert's "Ungeduld" ("Impatience"), for instance, seems to tumble over itself with delirious happiness, and carries irresponsibility on the very face of it. The singer's hat might blow away into the river and he would never know it was gone. That dishevelled mood he has to express — albeit not in his technique.

Charles Wood's "Ethiopia saluting the Colours" is a dramatic song relating an incident in the American Civil War. Though no doubt the incident took place, the two characters concerned are meant to be symbolical of slavery and emancipation. That comprehensive symbolism gives the atmosphere and indicates a mood of mystery.

Francis Korbay's "Mohàc's Field" is a Hungarian folk-song nominally expressing a man's heroic battle with ill-fortune; but the atmosphere is essentially one of *virility*, and the mood or attitude one of active combat. The singer will find at the end of this song, if he has been in the mood, that every muscle in his

body has been stretched taut in unconscious response
to the play of feeling.

All such physical response, and all facial expression,
should be unconscious and automatic; for the very
idea of artificiality is abhorrent. None the less, mood
and physical response are so interdependent that —
paradoxical though it may sound — the response can
sometimes actually appear to initiate the mood. If in
the song just quoted the singer will but tighten his
muscles and set his teeth before the first chord of the
opening symphony is played, he will find that he has
apparently thereby got himself into the mood. This,
however, is a variant not to be recommended to
beginners.

There are certain forms of mental expression,
concomitants of mood, which are so insistent as
to give the singer an impression of actual physical
demonstration. The shrug of the shoulders, for
instance, in

"Und er lässt es gehen alles, wie es will,"
"Little does he trouble, come whatever may,"

in Schubert's "Leiermann"; the leap back of the
knight in Schumann's "Waldesgespräch," in the
words:

"Jetzt kenn' ich dich,"
"I know thee now,"

as he recognises the Lorelei; or the collapse of the
"Laird of Cockpen" (Hubert Parry) on his refusal by
Mistress Jean. All such moods must be illustrated
practically by the voice alone. How this can be done
will be shown directly.

Most songs carry their atmosphere on the surface.

They tell their own tale, and the singer has but to follow the beaten track. But to the student they cannot compare for interest with those in which he has to look for it.

There is one thing that will help him in the search. Every song has a signpost hidden somewhere. The student, in the process of absorbing his song, will find that gradually, imperceptibly, one sentence or phrase, generally in music and words alike, will begin to stand out, to impress itself upon him as typical of the atmosphere of the whole, a guide to the whole mood. This sentence is the key to the song — the *master-phrase.* Every voice has a master-note which shows the character of the whole, and from which the voice can be trained up or down. Likewise every song has its master-phrase. Some songs, as said above, are their own master-phrase from sheer obviousness, but even when the atmosphere is most subtle the student hardly needs to look for it. The signpost will loom up through the darkness. The master-phrase will come of itself. Round it all the other sentences and phrases will group themselves and settle down, and order emerge out of chaos. It need not necessarily be emphasised in the actual singing : it is simply a master-phrase to the mood.

For instance, in Schubert's "Leiermann" the master-phrase is the sentence quoted above

"Und er lässt es gehen alles, wie es will."

The shrug of the shoulders referred to there is but the outward sign of the dreary hopeless indifference of the mood. If the singer can sing that phrase in that mood, it will colour the whole song.

c

In "Feldeinsamkeit" the obvious master-phrase is in the last line already referred to.

> "I feel as though I long were dead and borne along to heaven."

In some cases there are two master-phrases, either illustrating two moods, as in "Auf dem Kirchhofe" (see above) where the two words "gewesen" and "genesen" are the respective keys to their moods, the one *relentless*, the other *peaceful;* or, as in Stanford's "Fairy Lough," where the first master-phrase, "lies so high among the heather," gives a feeling of remoteness that makes the whole song sound far away, this being followed by a supplementary phrase, "and no one there to see," which tells why it is that the fairies are not afraid to show themselves.

This use of the master-phrase applies particularly to that class of songs called "Atmospheric," which will be dealt with further in the chapter on the Classification of Songs.

The principle has one other great advantage. It can be used therapeutically. A singer often finds that after the constant singing of any one song, that song seems to have lost its charm both for himself and his audience; its power seems to have become atrophied. If he will examine it closely he will discover that he has begun to over-elaborate his detail; the song, *as a whole,* has faded away, and the making of points has monopolised his interest. Such over-elaboration is generally the symptom of staleness. There is then only one thing to be done. He must give the song a rest. Let him put it away for some months, and then go back to it and look for his master-phrase. As if by magic

the atmosphere will grip him afresh and make the
old mood new.

In the whole treatment of Atmosphere — it sounds
sententious, but it must be said — the singer must be
possessed of a wholesome respect for truth.   Artificiality and Atmosphere are a horrid contradiction.
His individuality is shown in adapting his powers to
the atmosphere of the song, not in inventing new
moods to suit his own powers.   He has to ask himself
not what can be put into the song but what can be
got out of it.   If the music, for its part, does not tell
him more about the poem than he knew already, it is
not worth much.

COMMAND OF TONE-COLOUR

Hand in hand with Atmosphere goes Tone-colour.
Song-painting without it is impossible.   Whether it
is a pure gift or capable of acquirement, it is hard to
say.   It can be successfully imitated, no doubt, but
assimilated tone-colour bears somewhat the same relation to natural tone-colour that the *trillo di agilità*
bears to the *trillo di natura*.

Tone-colour is part of the physical *response of the
voice to the play of feeling*, and, being physical, is
correspondingly hard to describe in writing.   Every
individual has command of, and unconsciously uses,
a certain form of tone-colour every day of his life.
The ordinary man — he does not need to be a singer —
can, by the mere giving out of breath in a variety of
sighs, express in colour such varied emotions as sad
contemplation, surprise, pleasure, horror, contentment,
amusement and so on.   Tone-colour in singing is the

vitalisation of that breath before adding to it the spoken word. It follows the mood as unconsciously as the hand follows the eye or the sigh follows the thought. To successful interpretation it is indispensable. Without it variety or dramatic illustration would be impossible, for it is the singer's substitute for stage-setting and action.

Tone-colour is of two sorts, *Atmospheric*, in which the colour paints the mood (as described in the last chapter), and *Dramatic*, in which the voice adopts a character or series of characters, and stages, or illustrates, the actions and emotions of each. The one is passive, the other active. The stage-settings of "Feldeinsamkeit" and "Mohàc's Field" are as far apart as the poles; so are their tone-colours. The one drugged, half asleep, ready to accept any insult with an indulgent smile; the other tense, alert, virile, with clenched teeth and fists up.

In "Ein Ton" by Cornelius the voice-part consists of five sentences all sung on one and the same note. Here a man is thinking of a woman who is dead. Each sentence tells a different thought and a different emotion; without differentiated tone-colour every thought, no matter what the words, would *sound* alike. It is in reality a study of five emotions in five colours, a master-piece of unmonotonous monotony.

Take Schubert's "Erl-König." Here father, child and Erl King have to be clothed, staged and acted, each in turn and each differently, while the horse and the wind are staged or heard in the accompaniment. All five have this in common that *haste* and *fear*, either felt or inspired, run through the *different* tone-colours of each.

Or take Schumann's "Waldesgespräch," another variant of the "Erl-König" type. There are four colours here, two for each character. For the knight a *"preux Chevalier"* sound in the first verse and the first half of the third, changing suddenly to a hoarse horror as he recognises the Lorelei. For her, in the second verse, a subdued enticing far-away sound, with a suggestion of timidness and a feeling of romance as she speaks of the Waldhorn far off in the wood; then the other colour, cruel and concentrated, getting harder and harder as she comes closer and closer, telling him the while who she is, with a leap at his throat on the words "nimmer mehr."

For purposes of illustration nothing is better than the true traditional ballad. This is generally in strophic or stanza form and has many verses sung to the same tune. It tells a story or plays a little drama in song, and the dramatic singer acts it in tone-colour. Sung without tone-colour it may awaken a certain amount of interest merely as a story, but no more than if it were read in a book; probably less, for the reader would read a variety into it for himself, which the singer's monotony would actually counteract. All the instrumental devices in the accompaniment of the musical adaptation cannot galvanise the story into life, if the singer does not stage it with his voice.

Let us take the famous old Scottish ballad, "The Twa Sisters o' Binnorie" (arranged by Arthur Somervell), and work it out in dramatic form and see how it can be coloured. There are various versions of the tune, but the following one will do admirably:

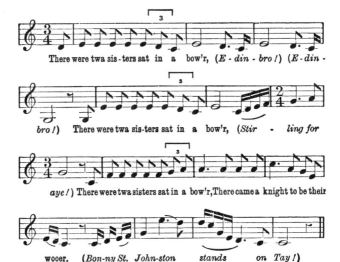

There were twa sis-ters sat in a bow'r, (*E - din - bro !*) (*E - din - bro !*)   There were twa sis-ters sat in a bow'r, (*Stir - ling for aye !*) There were twa sisters sat in a bow'r, There came a knight to be their wooer. (*Bon-ny St. John-ston stands on Tay !*)

In this ballad we have the following characters:

1. The eldest sister (jealous, and wicked for the dramatic purposes of the story).
2. The youngest sister (innocent and presumably beautiful).
3. The knight (deceitful, but not necessarily originally so).
4. The miller's son.
5. The miller.
6. The body of the drownèd woman (given a verse to itself and unquestionably treated with colour).
7. The harper.
8. The harp (a supernatural personality gifted with speech).

The story tells itself, but what a story and what

a chance for the singer! Eight characters, a dozen dramatic incidents and twenty-seven dramatic interjections! Surely there is scope for colour here; or are sisters, knight, harper, harp, *Edinbro', Stirling,* and *Bonny St. Johnston* all to sound alike? Let us take it line by line and act it with our voice. If the colour is there it will follow the mood.

THE TWA SISTERS O' BINNORIE

(Traditional.)

There were twa sisters sat in a bow'r,
  Edinbro', Edinbro',
There were twa sisters sat in a bow'r,
  Stirling for aye,
There were twa sisters sat in a bow'r,
There came a knight to be their wooer,
  Bonny St. Johnston stands on Tay.

Narrative. A plain statement of facts in ballad style. In practically strict time.

He courted the eldest wi' glove and ring,
  Edinbro', Edinbro',

Narrative, with a suspicion of contempt, and rather hard quality. Strict time.

But he lo'ed the youngest aboon a' thing,
  Stirling for aye,

Soft, and rather far away, to show the *secrecy.* Hold it back slightly and dwell gently on it.

The eldest she was vexed sair,
And sair envied her sister dear,
  Bonny St. Johnston stands on Tay.

Hard and concentrated, with a feeling of clenched teeth and tears kept back. Strict time.

She's ta'en her sister by the hand,
  Edinbro', Edinbro',
And down they went to the river strand,
  Stirling for aye,

Soft and concentrated. A suppressed "wolfishness" of tone, with a lilt in the rhythm to show them walking down to the river together swinging their clasped hands. Strict time.

The youngest stood upon a stane,
The eldest came and pushed her in,
  Bonny St. Johnston stands on Tay.

A rising horror as the action is illustrated; *stringendo* to the climax.

Sometimes she sank, sometimes she swam,
  Edinbro', Edinbro',
Till she came to the mouth o' yon mill-dam,
  Stirling for aye,

Breathless, fighting for life, hurried along by the stream.

And out then came the miller's son,
And saw the young maid swimming in,
  Bonny St. Johnston stands on Tay.

The same. The maid, not the miller's son, dominates the scene and colour.

"O father, father, draw your dam,"
Edinbro', Edinbro',
"For there's a mermaid or a swan,"
Stirling for aye,
The miller quickly drew his dam
And there he found a drownèd woman,
Bonny St. Johnston stands on Tay.

Breathless. Hurried. The miller's son *runs.*

Quickly.
Quickly up to the word "found." He does not see what it is till at the word "found." Then a perfectly different colour — very quiet to show that she is dead and still — then a slight pause as one sees him step back in horror and hears him whisper under his breath, "a drownèd woman!"

Round about her middle sma',
Edinbro', Edinbro',
There went a gowden girdle braw,
Stirling for aye.
All amang her yellow hair,
A string of pearls was twisted rare,
Bonny St. Johnston stands on Tay.

The whole of this verse uniform in colour. With a swinging lilt to show the beauty of her figure and the golden girdle, the long hair clinging round her body and the string o' pearls twisted in and out.

And by there cam' a harper fine,
Edinbro', Edinbro',
Harped to nobles when they dine,
Stirling for aye.

Straightforward. Pure narrative, but with a slightly concentrated tone, as though beginning to gather up the threads of the story and work towards the climax.

He's ta'en three locks of her yellow hair,
And with them strung his harp sae fair,
Bonny St. Johnston stands on Tay.

A sudden change to *piano* and a half-spoken tone. The supernatural element enters here for the first time. All possible beauty of tone in the last line to show "the harp sae fair."

He went unto her father's hall,
Edinbro', Edinbro',
And played his harp before them all,
Stirling for aye.

Narrative, but with the same concentrated tone and feeling of pushing-on to the *dénouement,* though without any increase of pace.

And soon the harp sang soft and clear,

Narrative, but "soft and clear." Every word must tell. Appreciably slower.

"Farewell my father and mother dear!"
Bonny St. Johnston stands on Tay.

Far away, unearthly, a spirit voice. Quite *ad lib.* in *tempo* provided that the phrase is only spread out and the rhythm kept intact.

And next when the harp began to sing,
Edinbro', Edinbro',

The same far-away deliberate sound. Narrative, but narrative dominated by the spirit-voice.

'Twas "Farewell Sweetheart," said the string,
Stirling for aye.

Still more remote, seeming to recede farther and farther away, getting slower and more distant right up to the end of "*Stirling for aye.*" Then a pause. Then —

And then as plain as plain could be,
"There stands my sister, wha murdered me!"
Bonny St. Johnston stands on Tay.

There is a pointed finger in the very lines. Sung "plain as plain could be" the words ring out clear. Every eye turns on the sister, the knight leaps to his feet, the music stops and the harp-strings break with a snap.

It will be noticed that in each case the italicised "tag" or interjection has been treated as part and parcel of the line immediately preceding it, and must, therefore, be given the same colour. These may have been once an accepted participation by the audience in the telling of the story, but nowadays they have to be sung by the soloist, and by him alone. In any case, whether those lines belong to either or both, they must help to tell the story and be coloured accordingly.

Tone-colour in such dramatic characterisation must be handled with care; otherwise it may develop into caricature. No man could convey, or even attempt to convey, the actual difference in pitch or *timbre* between a woman's voice and a man's, between the miller's son and the drowned woman of the harp, or the Knight and the Lorelei in "Waldesgespräch." Sex settles that for good and all. But interpretation knows no restriction of sex, and stages its individuals by differentiation in the handling of the character and sentiments of each.

The old English "Madam, will you walk?" is an admirable case in point. Here we have a little sparring-match between a man and woman, a miniature model for refinement of handling, sung *mezza-voce* almost throughout and led by gentle stages to its happy end. But the modern realist changes the scene from country lanes to city slums, and coarsens the demand for deference into outraged virtue and the want of it into vice.

It may be urged that the line by line treatment of the ballad is a direct contradiction in terms of the advice on Atmosphere to sing a song *as a whole*, and that such a handling must involve "patchiness" and over-elaboration of detail, the very faults most deprecated earlier. Strange to say, in the *accompanied* strophic ballad (the difference in the treatment of the accompanied and unaccompanied traditional song will be dealt with in the chapter on the Singing of Folksongs, p. 216) this does not apply; in fact, the exact opposite applies. The *accompanied* strophic ballad without detailed dramatic illustration is intolerable. The adapter has in the case of the "Twa Sisters" realised that fact to the full, and in his accompaniment has admirably illustrated the text verse by verse and line by line; the singer must do likewise. Curiously enough the impression left at the end on the mind of the listener is never the illustration of any line or lines but the atmosphere of the ballad as a whole. Like a dream a whole drama has been played in a moment or two of time, and like a dream its spell holds him when he wakes.

The modern composer when writing original music, to the ballad form of poem bears the danger of "patchiness" in mind. His setting is consequently seldom strophic (for the modern strophic ballad for some unaccountable reason is a poor thing beside the old) and is generally so dramatically varied to suit the action that the singer finds it all ready-made for him. Stanford's "La belle Dame sans Merci" is an excellent case in point. Here the whole story — a variant of the same old story of the Knight and the Lorelei — is so staged in the music that the singer has only to *feel* it and follow out the stage directions. But through the whole ballad, illustrated as it is line by line, runs the indescribable atmosphere of the spell —

> :"And this is why I sojourn here,
> Alone and palely loitering."

It is not merely a line by line ballad; it is an atmospheric drama, sung and acted in colour *as a whole*.

The student will ask with justice at the end of all this, "What is Tone-colour? What is the physical colour of each and every sentiment of each and every line of your ballad, and how am I to get it?" That can never be put in black and white any more than the colours of a picture.

Tone-colour is a hybrid word partly borrowed from the sister art. It is impossible to say of any character or line in the above ballad that it is red, white or blue. The utmost the singer can do is to *vitalise the breath* and add to it the same colour when *singing* the words that he would give when *speaking* them in accordance with their dramatic significance. How that colour is *physically* to be attained is a matter for his teacher first and himself later. It can never be put on paper. Though physical in its process and actual in its acoustic effect, it makes in reality a moral appeal to the listener. It is, as said above, the unconscious response of the voice to the play of feeling, and is unconsciously assimilated by the sympathy of the hearer. It is only in the big dramatic changes that its physical side becomes apparent; in the rest it is too subtle, too psychological in its appeal to gain actual recognition. It is for this reason hard to imitate and inclined to be classed among the gifts. One thing is certain about it — *it means power*. It holds attention and is Interpretation's best friend.

STYLE

THERE is style in the doing of everything under the sun; in the wearing of clothes or the sailing of a boat, in the swath of a scythe or the lilt of a song. It is the backbone of every great game, and the heart of all the arts.

Style in singing is popularly associated with the finished performer. The public has never given it a name — only the expert does that — it has simply felt it, chuckled over it, and taken off its hat to the old hand and the "grand manner." The public is right, though the wrong way round; it thinks the singer has made the style, whereas, in nine cases out of ten, the style, lovingly handled, has made the singer. Were style but the result of competent treatment of material, the obeying of rules and hard work, the world of song would be a world of beauty. Were it the exclusive property of the old hand, its place in this book would be at the end; but style is essentially part of "Equipment," though for definite reasons it has to be classed by itself.

Some singers have no sense of style or anything remotely approaching it; they may recognise it in others dimly when they see it, but they invariably attribute its effect to some physical or technical feat of the performer.

29

Others have style and nothing else. Nature sometimes has her tongue in her cheek and turns out freaks, and foremost among them is the sterilised stylist. In rare instances he has not even intelligence; but generally he fails in will-power. His prototype elsewhere is well known. He is a commonplace in cricket. He has been played again and again by his puzzled captain for his school or his club, and has never come off; dropped at last, he goes home sour and disappointed, a grumbler for life. He has not remembered the Parable of the Talents — he has not learned his business. He has lived on his capital, borrowed from his friends and ended in the workhouse. Style is capital to invest, and style in singing is a fortune held in trust for music; the man who neglects it is not only a bad economist but a fraudulent trustee.

The singer who has not got it to begin with can often by hard work and laborious imitation achieve a fair substitute for it, though it is quite distinguishable from the real thing. All honour to him when he does. If he is made of that fine stuff, let him see to it that his models are worthy of his will-power. Tradition is his best friend, and the man and the readings who have stood the test of time will be his stand-by.

True style is born, not made. Like magnetism, it is psychological. Why should one man have it and nothing else? Why should another have all the other gifts and have that left out? Why should one child the first time he takes a bat in his hand make the old cricketer's heart leap, and another leave him cold? Why should two men sing the same song

note for note alike, and one move his audience to
tears, and the other to yawns? Is it style or
magnetism, or both? It is an extraordinary fact that
style and magnetism are generally found together.
They hunt in couples, but which leads and which
follows none can tell. The magnetic singer is so responsive to magnetism that instinctively his faculties
may meet the call, and style may be the result; this
may account for the quasi-success of the "style-alone"
man, spoken of above, who has not learnt his business.
But, on the other hand, style will not play second
fiddle to anybody; so compelling is he in his very
presence that when he is there the listener, be he
amateur or expert, throws up his hands and begs him
with a smile to rifle his pockets.

Those who have had the privilege of hearing Sir
Charles Santley sing, say, Gounod's "Maid of Athens,"
have felt its spell. A straightforward, simple, little
song, but made into a masterpiece by the hand of the
magician, word by word, phrase by phrase, line by line,
perfect in balance, virile in appeal, with no cheap
pauses, no sentimental cadences, it runs to its inevitable
end. Born style, fostered in childhood, trained like
a Spartan and delighting like a giant in its strength,
takes song, singer and hearer in its arms, and tumbles
them out together head over heels, hatless, coatless
and rejoicing. No wonder the Maid of Athens was
in no hurry to give the poet back his heart!

Every song has a style of its own suited to it;
but this chapter treats of Style in its general sense
comprising them all. How to define it is the difficulty.
It is the *treatment of the subject "in large,"* both in
conception, phrasing and colour; the turning out

of a work of art in which the component parts fit
in in proper proportion in the right places, and are
forgotten in detail; where no single phrase *undesignedly*
attracts notice to the detriment of its neighbours;
where commonness or smallness are as conspicuous
by their absence as cheapness of effect; with a quasi-
aristocratic air over all, leaving the listener with the
feeling that the reading was as *inevitable* as it was
true.   This is naturally a counsel of perfection,
something to be aimed at; but style is a powerful
stimulant.

In two things the master of style will never fail —
he will treat his song as a whole, and rhythm will be
in his very system.   Rhythm is the heart of music,
and lilt the lungs of style.   If those two are sound
he is ready for anything, from the 120 hurdles to a
Marathon race.

Style like Magnetism is surrounded by enemies.
Some of these will be discussed in Part II., but the
three most deadly may be treated here.

Its first great foe is *cheap effect*.   Cheap effect is a
poisonous germ carried in the blood, with which the
friendly phagocyte of taste is ever at war; if ever it
gets the upper hand, good-bye to the body politic !   So
opposed is cheap effect to style, that the man who
knows the one may be said to be ignorant of the
other.   There are some singers whose eyes are always
on the audience, not as ambassadors of magnetism, but
as conscious seekers for effect.   Such singers can never
have style or even achieve the singing of a song as a
whole.   Their attention is so concentrated on detail
and its handling for cheap effect, that the song in large
is lost, and with it Atmosphere, Mood and Style.

The next is *over-elaboration of detail*. This comes far oftener from over-familiarity than from deliberate intention (*vide* p. 18 above). The mind and ear become so accustomed to the work that, if the performance is not to grow purely mechanical, new readings must be given to the detail; the individual phrases are over-emphasised or distorted, the natural proportions are lost, the work loses its balance and the sense of style disappears. The performer does it to satisfy himself not his audience, but its effect is the same — you cannot see the wood for the trees. It is a commonplace in all branches of executive music, conducting, pianoforte-playing or singing. The only remedy for unconscious over-elaboration is rest. The work should be given a holiday. Conscious, though involuntary, over-elabora-tion is generally the fault of the beginner, whose mind has been so obsessed with technique that it has not had time to grasp the song in large. If the beginner is made of the right stuff, it soon disappears. In some cases it is always present, but then style is always absent.

Deliberate over-elaboration stands on a different footing. It is at daggers drawn with style, and with full knowledge jumbles up values and destroys per-spective. It is the mother of mannerism and the stock-in-trade of burlesque. The following quotation from a recent illuminating article in the *Times* on "Mannerisms" puts it in a nutshell : [1]

"It is when we look at the presentation alone that we find that, as the perfection of workmanship decreases, so does the obtrusion of mannerisms

[1] By kind permission of the writer of the article and the Editor of the *Times*.

D

increase; for perfect workmanship is workmanship become instinctive, and the insertion of mannerisms is the result of conscious manipulation. With the giant, style includes incidental mannerisms; with the pygmies, the mannerisms constitute the style. There are, of course, little turns that one expects to find, even with the giants. But our point is that, with the really great, idiosyncrasies have subconsciously become a vital characteristic of style, and we merely feel that the personality of the creator presupposes certain lines of action. The greater the man, the more difficult it becomes to place your finger on a square inch of his music and say, 'he was always working off this little trick'; and it becomes difficult almost to the point of impossibility to say, 'he reverted consciously to this trick because his inspiration ran dry.' Consequently the great men are extraordinarily difficult to parody; for the parody either falls flat as mere reproduction, or it shows its hopelessness by becoming, under the inspiration of its model, something uncommonly like music.

"But with the smaller men, those to whom technique has never become second nature, but whose characteristic mannerisms are self-consciously dragged in to conjure up the idea of personality — with these even an unskilful parodist may acquire an easy reputation. They are exasperating for so many reasons; for they are in the first place flaws, and they are also signposts which the unwary recognise with joy and gladness, and they further create the suspicion that the composer has said in his laziness, 'This is sufficiently like me to do for that bar,' without considering whether that bar had any justifiable place in the general scheme. But this last reason, amounting as it does to a charge of artistic dishonesty, opens up the whole question of inspiration, or continuity of thought, as opposed to manufacture, or conscious construction."

This, of course, was written of the composer, but *mutatis mutandis* every word of it applies to the singer and his methods. He does not need to be a giant or a great man; merely an individualist and master of his will. Whether the cause be artistic dishonesty or playing to the gallery or mere narrowness of vision, the effect is the same on Style.

The third and the greatest enemy not only to Style but to Magnetism and every branch of Interpretation is *Self-consciousness*, unfortunately a commonplace of the English singer. Englishmen are noted for their reserve, but reserve in public performance is two-sided. It may either, as the enemy of exaggeration, be the friend of style or in its self-conscious form go hand in hand with monotony. Metaphorically speaking, the average English singer sings in the high collar of caricature. Armoured in reserve to his ears he cannot look up for fear of hurting the back of his neck, and cannot look down for fear of seriously damaging his chin. With a magnificent spirit of compromise he looks straight ahead along the level road of mediocrity, breathing convention through either nostril. He cannot be altogether blamed. Angels in triplets and organ obbligatos do not demand a wide range of vision; familiarity accepts the old friends blindfold, and though the top A on "Heav'n" may move the singer to apoplexy and the audience to hysteria, it leaves the collar intact. O for a fire to cremate the starched collar, and a rude north wind to scatter the cheap conventions !

The self-conscious singer cannot forget his *Technique;* he cannot forget his details — his mind cannot travel far enough away.

He cannot forget himself; he cannot, therefore, give and receive *Magnetism.*

He cannot visualise; he cannot let his imagination run; he cannot, therefore, feel *Atmosphere;* he cannot find the *Mood.*

His voice cannot *unconsciously* respond to the play of feeling; he cannot, therefore, paint in *Tone-colour.*

He cannot think of his song in large; he cannot, therefore, have *Style.*

The singer has only two things to think of, his song and his audience, and of these the song comes first, and a long way behind — a very long way — comes the audience. That precedence is due not only to the song but to the audience as well, and well the audience knows it. The better the work of art, the better the value given, and the greater the respect paid to the buyer. The singers who assign precedence to song and audience respectively are wide as the poles apart. The cheap appeal to the audience is that most generally given at the expense of the art if not of the song. It is the friend of self-consciousness and the deadly enemy of style. Style and self-consciousness are like Jekyll and Hyde. When the one steps in, the other shuffles out ashamed.

PART II

RULES

THERE are certain rules which apply to every song in existence. They must have been so assimilated into the singer's very being as to be forgotten in detail and to become unconscious in their application. Their observance is not only due to the song as music, but is essential to successful performance, for they are intimately associated with the knowledge that singer and audience sing, in reality, together in sympathy. There are three main rules — few but comprehensive — and of these the first is *musically* far the most important, for it is the mainspring of all singing, from a phrase to a song-cycle.

MAIN RULE I

NEVER STOP THE *MARCH* OF A SONG

"Music do I hear?
Ha! ha! keep time: — How sour sweet music is,
When time is broke, and no proportion kept!"
King Richard II.

Every song *marches;* it moves in companies of unit notes from point to point and marches *in step* to its appointed destination. It can march slower or quicker or halt *at command,* but no unit may stop on his own account.   The shoulder-swing of the marching regiment is the lilt of the song.

There are various reasons for, and ways of, stopping the march of the song: over-elaboration of detail; pauses for cheap effect; even deficiency in rhythmical sense, though the man without rhythm does not get very far in public.   But in nine cases out of ten the reason is physical — want of breath.

The average singer, when he finds himself at the end of his breath, adds an extra silent beat to the bar — in some glaring instances more than one — in order to refill his lungs with the least possible physical discomfort to himself.   When this is done in every other bar, time-signatures become a farce, and rhythm, the beginning of all things, relapses into chaos.

The man who has no sense of rhythm has no right to inflict himself upon the public.   The man who has it and lets it go because he cannot hold it, has not learnt his technique; he must go back to the beginning and learn the first essential of his business — how to breathe.   The man who has it and deliberately destroys it, is guilty of murder in the first degree.   It is the most heartless, sacrilegious, slipshod crime in music; it is like strangling a child, for from rhythm music grows to manhood.

This does not mean that music to be rhythmic must be foursquare.   Heaven forbid !   The child must have playtime.   Any phrase or set of phrases may be spread out or narrowed in, held back or hurried on, and the rhythm will be all the better for the change when it is brought back.   The stride may lengthen or shorten, but the marcher must never get out of step.

The motor-driver knows the sensation of the recurrent missfire, the little kick which holds back the run of the car for a fraction of a second. That little kick gets on his nerves in time. The missfire in the song — due to some little defect in the running of the engine — gets on the nerves of the audience; do not singer and audience sing or cease to sing together? As long as his singing is rhythmically continuous they stay with him on the level of his song. Let him *stop* that song to take breath, and the magnetic thread by which he holds them snaps and down they fall; that means starting again, and it must be a matter of many bars before even the most magnetic performer can wind them back to his level. Let him repeat this once or twice, and by the end of his song they will have been left hopelessly behind, having given up the struggle as a bad job. It is not a matter of aesthetic interpretation; it is a primary physical response to the demand — positive though unexpressed — of the audience for fair play. If the singer halts, they halt; if the song flags, so does their interest; if the leader stops his march to rub his knee, the man immediately behind him jostles into him and, like the motorist, probably swears.

This first rule is a law of the Medes and Persians. To keep it is the hardest thing, physically, in singing. It involves Spartan training, bodily strain and great courage, for it is for years one long struggle between mind and body. Nature cries aloud for rest, for breathing-time; but it should be a point of honour never to give in. If the singer realised the power of rhythm to sway his hearers, he would never leave it till he was its master. The rhythmic drumming of feet of the popular audience may sound cheap, but it is music to the singer's ears; it is a tribute to his rhythmic sense; it means that the lilt of his song has found his audience out. The people hungers for rhythm, but in this country its caterers either leave it to starve or adulterate its cheap food.

There are rhythmical and unrhythmical countries — those who have heard a waltz played in Vienna and London respectively, will appreciate the difference — nations in whose blood rhythm is born, and others where it has no home, where, when it comes as a visitor, it is welcomed with delighted surprise. It is only within the last few years that rhythm has been recognised and taught in our elementary schools as the first essential of musical training; the effect is being widely felt already, but there is still a long way to go. The average Englishman, and even the average trained English musician, associates rhythm with primary and secondary accents alone, and lets the rest take care, more or less, of themselves (to the average singer $\frac{6}{8}$ is the only compelling rhythmical time). If he religiously accentuates the first and third beat of the four-beat bar, he feels he has done his duty. If he had ever heard an Irish piper and fiddler play a dance-tune, or seen and heard a troupe of Spanish dancers, he would suddenly discover that there is an accent on *every* beat. Let him get some violinist blessed with a strong sense of rhythm to play him the following well-known Irish tunes:

"THE FLANNEL JACKET."

"WHO'LL COME FIGHT IN THE SNOW?"

Con spirito.   (From DR. ALFRED MOFFAT'S COLLECTION.)

or the "Zapateado" (Sarasate), with an accent on *every* beat, and he will find, not only that the primary and secondary accents are there as before, but that some extraordinary thing has happened to the tune and to him; that the rhythm has suddenly ceased to be vertical and has become horizontal, that the pendulum no longer swings up and down in the old gentlemanly way; that some great undiscovered elemental force has caught him in its mighty grip and is whirling him along *in a straight line* to the inevitable end.

True rhythm is inexorable; true rhythm is compelling; true rhythm is ever on the move and ever in a straight line. Nothing can stand before it; everything must clear out of its way. Its motto is "Push on!" No singer could, or should, sing such instrumental accents as these, but the feeling of their rhythm should be in his blood, and "Push on!" should be written in letters of fire in his brain, for it is the secret of the singing of every song, big or little, fast or slow; be it as harassed as the Erlking, or as lazy as Feldeinsamkeit, it pushes on *in a straight line* to its goal, *inevitably*. It is this principle of the straight line which makes fine phrasing, and the sense of inevitableness which gives the impression of style. To the singer — who is, as a rule, the least musical of musicians — accents are associated with down beats. He knows, of course, that the third or secondary beat of a four-beat bar is horizontal in conducting, but none the less in his mind's eye all accents are descending hammers vertical in direction. If this were so a song would never move; it would be a series of accented and unaccented notes jogging up and down. But rhythm is like a piston; it may apparently work up and down in direction, but it drives the structure *forward*. Melody is horizontal, and melody is ever on the move and on the move forward; and melody is the singer's stock-in-trade; if the melody wants to move, he must move with it or for ever hold his peace.

A song must "push on" note by note, word by word, phrase by phrase, to its inevitable end.

Phrasing.

The musical phrase is like a wave. It may move in a continuous *crescendo*, gathering force as it goes, and dash itself against the rocks; it may recede *diminuendo* and disappear; it may join forces with its neighbour for the common good, or break away for its own ends; or it may follow its leader in pure *cantilena* ripples to the shore. Occasionally it lands in some quiet pool and seems to die — such dead levels in song are used for special purposes of colour-illustration and contrast. The natural phrase has motion inherent in it, and whether its wave be salt or fresh, self-sufficing or subdivided, rising or receding, breaker, roller, or ripple, it moves in undulations irresistibly forward.

The singer's technical mastery of phrasing has been taken for granted for the purposes of this book; its early stages have been taught him by his master and the latter he is supposed to have learned for himself. There are, however, certain facts about phrasing in general which are of supreme importance.

As the strength of a chain is its weakest link, so the strength of a phrase is its weakest note (or absence of note) and the strength of a song is its weakest phrase. If the singer is physically in bad voice he will probably sing a bad note here and there in the course of a song; each bad note thus destroys the beauty and balance of the phrase to which it belongs; each phrase thus marred has in turn marred

the song *as a whole*. This is a thing the singer
cannot help, as it is probably out of his control. What
lifetimes of misery has not the public singer lived
through — perhaps producing a new work at some great
musical festival — when his voice has played him false,
and when in the fight with technique interpretation
has been knocked over the ropes!

Let him, on the other hand, be in splendid voice
and rejoicing in his strength; the temptation must
occasionally be irresistible to make effects — with his
voice, but at the expense of the phrase and consequently of the song. The mere joy of sound is too
much for him, and for the sake of it he stops the
march. Or again, the conscious pause — that most
dangerous of all the weapons in his armoury — may
tempt him. It is generally better left alone. Its
dangers will be pointed out later.

The phrase is surrounded by enemies; it does not
take a great amount of imagination to recognise our
old friends, cheap effect, over-elaboration and self-
consciousness, as directing the operations.

The singer would do well to bear this in mind.
In nine cases out of ten, where the music is good, the
phrase in itself is far stronger than anything he can
read into it. (In folk-songs in 99 cases out of 100.)
Phrases and phrasing depend upon structure and
balance. They run horizontally along a straight line
in a succession of curves (*more Hibernico*) and they are
always moving. Phrase is balanced by phrase, like
the wings of a bird. If the singer brings in false
values or pauses, he makes the movement vertical or
stops the march of the song; he puts a broken bone
in one wing and upsets the balance of the flight.

Alongside the great giants, Rhythm and Motion, human "effects" are pygmies.

Thus in the "Erlking" as soon as the Erlking begins to speak there is an undeniable temptation to slacken the time. A new character is being acted and a new tone-colour used; it is sung *piano* and with a certain grim *oiliness* directly opposed to the terror and furious heart-beating of what has gone before. The singer's first impulse is to make the change complete even to the matter of *tempo*, but, as said above (p. 20), through the whole song run *haste* and *fear*. The Erlking does not need to stop the horse to whisper in the child's ear; the singer does not need to stop the march of the song. This phrase or that may be slightly broadened (and brought back) — the horse may lengthen its stride — but song and horse *push on* at a gallop, in *haste* and *fear*, to the *inevitable end* — home. The bare idea of a singer's *rubato ad lib.* in the "Erlking" would make Schubert turn in his grave.

All the great "moving" songs are full of these pitfalls; Brahms's "Meine Liebe ist grün," Schubert's "Ungeduld," Schumann's "Widmung," and hosts of others, in which the rush of the song is its essential strength, are ruined by attempts to read "vocal effect" improvements into the rhythm. Take, again, Brahms's "Vergebliches Ständchen." Here the composer's general directions are "Lebhaft und gut gelaunt," *i.e.* "Lively and good-humouredly." There is not one *rit.* or *accel.*, or any direction as to change of pace, throughout the entire song, until just before the last verse, where it is marked "Lebhafter," *i.e.* "more lively." Yet the contralto has seldom been found who could resist the temptation, in her anxiety to stage the characters, of rolling the phrases round her tongue, juggling with the rhythm, reading in long notes and pauses, and stopping the forward movement of the song. As though anything she could drag into it from outside could approach the adorable fascination of the little rhythmic tune and its bubbling accompaniment!

There is another point in this question of balance which is continually arising. Many phrases in the voice-part are balanced by repetitions or answers in the accompaniment. This form of musical dialogue is one of the most delightful things in song. If the singer reads in false effects, the accompanist in his answer must either do the same or spoil the balance. It is a choice of two evils; for though by imitating the singer he may preserve the temporary balance of the phrase, the overloaded phrase will in its turn upset the balance of the song. Schubert's "Morgengruss" ("Die schöne Müllerin," No. 8) is a case in point. In the following passage the accompaniment repeats the singer's phrase:

muss ich wie-der ge - hen, wie-der ge - hen.

Suppose that the singer for purposes of vocal effect puts a *fermata,* or *tenuto,* on the word "muss"; there are then two alternatives. Either the accompanist can take no notice and play his repetition of the phrase as it is written, in which case the two phrases do not balance; or he can imitate the singer by also putting a *fermata* on the E, in which case the singer will have to abnormally lengthen the word "gehen" in order to give him time. (If the singer forgets to do it, as is probable, the accompanist will have to make a wild scramble to catch him up.) He will probably choose the first; but either of the two is sufficient to reduce the rhythm to chaos.

There are certain types of accompaniment which are positive danger-traps for casual *rubato.* They catch and hold the unrhythmic singer like bird-lime, and it is a bedraggled object that struggles out at the end.

Arpeggios, from their natural forward movement, cry out against being stopped. They shout "Come on!" in every note. There is no broader effect than the arpeggio broadened *colla voce,* but it must be broadened *as a whole*, not played one half *a tempo* and the other anyhow, *and it should be rehearsed.* The look in the eye of the accompanist, cut off in the middle of his *arpeggio* stride for a high note effect, is more expressive than a volume of sermons. Take the handling of Mendelssohn's "Auf Flügeln des Gesanges" or Schumann's "Der Nussbaum." In the first of these the composer has not marked a single *rallentando;* the "wings" of song do not require it. The charm of the song lies in the pure *legato* singing in a continuous dreamy straight line, and does not require any outside effect. In the Schumann song the composer has marked his *rallentandos* where he wants them; any gratuitous interpolation of effects not only is superfluous, but *sounds* laboured, and stops the song.

Accompaniments with a definite swinging lilt of their own are a danger, more from their own virtues than the singer's vices. In the following bars from Arthur Somervell's "Birds in the High Hall-Garden" ("Maud"), the impulse to call "Maud" on a series of lengthened note-values is strong; but the lilt of the accompaniment will not permit of it. There are many opportunities in the instrumental interludes of spreading out and bringing back the rhythm, but in the voice-part the composer has not marked a single change of *tempo.* The breeze playing round the cawing rooks gently swings the tree-tops; it does not blow in gusts. The singer can convey the sense of distance and reminiscence by tone-colour. Here, as elsewhere, his imagination will initiate and his voice obey; it will respond unconsciously to the play of feeling.

The same applies to accompaniments with a definite
rhythmic meaning. In Schubert's "Das Wandern"
the swishing of the waterwheel is the secret of
the song; and it is in the accompaniment. The
millwheel turns on and on at the same pace;
a *fermata*, or a *rallentando*, or even a *rubato* anywhere in the song would be an outrage. Take the
following bars from Stanford's "Johneen" ("An Irish
Idyll"):

E

Here the rhythmic figure in the accompaniment represents the rowing of a boat. A pause on the D flat on "duck" would make Johneen catch a crab. When the composer wants a *rallentando* he gives it, a few bars later, on the words "but the ship she must wait a wee while yet I hope." Here there is not the smallest objection to a slight pause on the word "wait," for not only does the word justify it as illustration, but the accompaniment shows that the boat is slowing up to the bank.

In the same composer's "Drake's Drum" the rhythmic figure is so obvious that the thought of juggling with it seems ridiculous. The old sea-captain is walking up and down his deck in Plymouth harbour and *thinking* of Drake. There is nothing in song or accompaniment to stop his walk, yet the amateur's tendency when he comes to the phrase, "Captain, art thou sleeping down below?" is invariably to sing the phrase twice as slow, and to pause on the *Cap* of "Captain" (and probably again on *down*), as though he stopped in his stride and put his ear to the deck, or tried to call up Drake on the telephone. The old sailor never *stops* his stride (D minor); but as he thinks of Drake and all he did and all he stands for, he *lengthens* it (*largamente*); his hands come out of his pockets, his head goes up and his eye lights with the joy of battle (D major).

In all such songs the accompaniment dominates the situation; the same applies to atmospheric songs, where the accompaniment has the illustration and the voice merely gives the mood. In all of these the singer's only effects are effects of colour; in matters of *tempo* he is subordinate. He must take his place in the scheme and march with it, even though the ascending phrase and the high note tempt him to look at the view when he gets to the top.

For some unaccountable reason syncopated accompaniments, which from their very rhythmic impatience cry out "Push on!" seem to invite dragging of the *tempo*. The singer's attention appears to wander from his own rhythm and to be absorbed in waiting for the other, and the tendency is to get slower and slower. Brahms's "Sapphische Ode" is a veritable contralto-trap. Here is an average contralto version of a bar or two:

⌢ = long pause for breath.

⌢ = long pause for breath.

Strange to say, with all its impatience, there is nothing more peaceful than the gentle syncopated accompaniment with its large possibilities of expressive *colla-voce* work. (Vaughan Williams's "Silent Noon" is a perfect example.) It is perhaps this very elasticity which betrays the contralto.

There is one other type of accompaniment which from its primitive simplicity and obvious adaptability to anything, seems to invite every fault in song.

The old series of repeated chords, in the following style:

forms the main accompaniment of three-quarters of all British so-called "ballads." It is the *vade-mecum* of the popular composer, and the old, old friend of the sloppy sentimentalist. No doorstep on a winter's night is complete without it. It has discovered more orphans than the combined force of the Metropolitan Police; it has saved more children's lives than the whole of the Country Holiday Fund.

In its extended triplet form

it is the only authorised ladder to heaven; the self-
respecting Organ Obbligato would faint at the suggestion

of supplementing any other style; it can play Ercles rarely, or roar you like any sucking dove; it is the embodiment of self-satisfaction, but is capable of fierce passion. Its love is like anything from a red, red rose to a Tannhäuser Venus. It has no particular drive in its rhythm, no imitation of the voice, no melodic figure, no atmospheric suggestion — just a good roast-beef, up-and-down, accommodating set of plain chords. Does the tenor want to stay in "Heaven" on that top A?— it is delighted to oblige; does he wish to hurry over the ineffective middle notes? — nothing could give it greater pleasure than to hurry along after him. He has the melody, it has the harmony; what more can anyone want? Rhythm? Balance? What will you ask for next?

How beautiful this accompaniment can be in the hands of a master, Schumann has shown us in "Du bist wie eine Blume," and Schubert in "An die Musik."

These do not pretend to exhaust the types, but enough has been said to show the *rhythmical* interdependence of voice and accompaniment.

In the art-song of to-day the singer has very little excuse to go wrong. The modern composer is such a master of prosody and metre, so eclectic in his literary choice of text, so conscious of the value of the singer as interpreter, that the latter finds his song ready-phrased and humanly expressed for him. Unlike the old days, when the human voice was looked upon more or less as a musical instrument of peculiar beauty and composed for accordingly, nowadays that voice is accepted as the most direct and powerful means for emotional ends, and the composer writing from his

heart the language of music trusts his interpreter to
tell it truly to the stranger in his own way. There
were giants of emotional expression in the old
days too — Bach was the greatest of all — still the
legacies of those times are for the most part dis-
tinctively musical rather than directly emotional in
their appeal. They belong largely to the three great
groups — florid, pure *bel canto* and rhythmical. These
three in their turn belong to the generic *bel canto*
class, in that they depend for their effect upon
perfected technique, beauty of tone and mastery of
phrasing rather than upon any direct aesthetic appeal
to the human emotions through illustrative expression
of the text. This is true even of Bach. In the St.
Matthew Passion we find on the one hand the recitative for bass No. 74, "'Twas in the cool of eventide,"
one of the most simply expressed and deeply moving
pieces of pure emotion in music, and on the other the
bass aria No. 51, "Give, O give me back my Lord,"
which, though moving enough when beautifully sung,
is yet so instrumental and quasi-florid in type as to
have to trust for its effect to beauty of performance
rather than direct expression of the words. The same
applies to the other bass aria, "Come, blessed Cross,"
or to the alto aria (No. 48), "Have mercy, O Lord,"
or the soprano (No. 19), "Jesus Saviour." Such arias
are contemplative in feeling, and have no direct
bearing on the great story, but by their very contrast
and passivity they intensify the active drama. Like
the Greek choruses they wrap us round with the
atmosphere of tragedy; as such we love them and are
stirred by them profoundly.

To all these arias and to all such florid arias this

Rule No. I. applies with all its force. They are essentially musical and largely instrumental in style, and in some cases have instrumental *obbligati*. The violin or the oboe will not *stop* the aria for bowing or breathing; no more must the singer. In some cases it seems almost beyond his physical powers to obey. The following passage in the tenor solo from Bach's church cantata, "Meine Seele rühmt und preist," looks frankly impossible:

Poco adagio (♩=60).

ist in meinem Gott er-freut

We know that it was written for the famous tenor of the Thomas-Kirche, though we do not know his name; but did he sing it as it is written, or did Bach write it simply as a delightful piece of music without consideration for its physical difficulties, and leave it to his trusted friend (or enemy) to phrase in his own way? If the former, the singer deserves a place in the Westminster Abbey of his country; if the latter, how did he phrase it? Was he, by the custom of his day, entitled to take a breath in the middle of the word (say at X and XX), treating the word merely as a vehicle of sound, and like an oboist breaking and

resuming the phrase without loss of rhythm or forward
movement; or was he supposed to re-arrange the
words

in   mein-em   Gott   .   .   er - freut, etc.

so as to give a semblance of language values? No doubt
the former, since the passage, being instrumental in
character, would, from the absence of complications of
diction at the breaking-places, thereby run less risk of
being stopped in its flow.

One cannot take the florid aria seriously as a direct
expression of its text. Florid passages occur which
are certainly admirably illustrative of a word or
sentence — the modern composer makes full use of
them *en miniature,* or in the form of the *melisma,* for
that very purpose — but the regular out-and-out florid
aria as a means of aesthetic expression is an anachron-
ism. Even "Rejoice greatly," from Handel's *Messiah,*
though far more plausible than most as an illustration
of the statement, is unquestionably a florid *tour de
force;* in fact, there is nothing much more technically
difficult to sustain than the famous passage

Re-joice   .   .   .   .

.   .   .   . greatly.

Only the best and most highly trained soprano can be
trusted to sing that passage well.

Next in difficulty — if not actually first — comes the pure *bel canto* song. The florid song depends upon economy of breath — upon the embroidering of notes on the smallest column of air sufficient for the purpose; it is generally fast in *tempo*, and its very pace hustles the singer forward. The pure *bel canto* song depends also upon economy of breath — upon the employment of the smallest amount of breath consistent with *sostenuto* singing, and its incidental *crescendos* and *diminuendos;* it is generally slow in *tempo*, and the slower the *tempo*, the more insidious the temptation to drag. Never in vain is its net spread in sight of any contralto. She feels her breath coming to an end, and says to herself, "This aria is so slow that a beat more or less won't be noticed!" so she takes her time (and the song's), breathes comfortably and stops the lilt, and down tumbles the audience. Laboriously she starts again, and laboriously the audience toils after her; in a couple of bars' time she repeats the process, and the audience, worn out with the effort to assimilate $\frac{5}{4}$ and $\frac{3}{4}$ bars in a $\frac{4}{4}$ song, like an autumn bluebottle drops from the window-pane dead. The names of such arias are legion. Giordani's "Caro mio ben," Handel's "Lascia ch'io pianga," *et hoc genus omne.* Mendelssohn's "O rest in the Lord," marked $\decrescendo = 72$ and therefore *andante* (if not *andante con moto*) in pace, is invariably sung *adagio*, sentimentalised, and dragged to such an extent that the song musically loses all motion and the "rest" degenerates into stupor. It is no wonder that the "Sapphische Ode" of Brahms spoken of earlier, pure *bel canto* as it is in style, when complicated with modern expression and syncopated accompaniment swallows its victims wholesale. Not only contraltos but all voices have to be ever on guard in *bel canto*. Whether it be in "Angels ever bright and fair" or "How willing my paternal love," the temptation to drag and to take illegitimate breathing-time seems to obsess singers, one and all. They forget that the slower the pace the less the power of the rhythm to take command, and that therefore they owe it all the greater loyalty.

Finally, there is the *purely rhythmical* song — that is, the song that depends upon rhythm for its effect, in which, in other words, the rhythmical flow dominates the situation, and all other forms of expression are subordinate; tone-colour, emphasis and diction, all may contribute or enhance, but rhythm is pre-eminent. The voice is here again looked on as a *musical* means to an end and has to make its effects by pure phrasing. Pure phrasing does not admit of false pauses to take breath; therefore, like the *florid* and *bel canto*, the *rhythmical* song must never stop its march.

Rhythmical songs are countless in number; this is not to be wondered at seeing that rhythm is the protoplasm of music. They exist in hundreds, written for every voice, from Arne's "Where the bee sucks" to "La danza" of Rossini. They may, as in both these cases, be admirably descriptive of the subject of the song, or as in Scarlatti's "Già il sole dal Gange" simply a delightful tune, applicable to a dozen other lively emotions; but they are one and all dependent for their effect upon the rhythm being kept intact. Because Ariel, being a thing of beauty, hopped by nature *rhythmically* from cowslip bell to bat's back, or because a dance is by nature rhythmic, we do not therefore consider either one or the other, any more than the sunrise on the

Ganges, as essential to rhythm; whereas rhythm is essential either to them or the song, or both. There are quantities of similar modern examples, some of which will be referred to later on, but at this stage we are dealing with the older types. To every single one of them, old or new, Main Rule I. rigidly applies.

Now in all these three, florid, *bel canto* and rhythmical, the danger-point is the compulsory taking of breath in places where no breathing-time has been provided by the composer. The singer must be so physically trained that he can take that breath at lightning speed, and must have so absorbed the first rule of phrasing that its application has become unconscious. That rule is, that wherever breath has to be taken *in spite of* a phrase — *i.e.* where no pause is marked — *the time-value must be taken from the note that is left, not the note that is approached;* it is the place you land on, not the place you take off from, which matters. To do this the muscular control must be supreme, for not only must breath be taken in a fraction of a second, but one note must be left and the other attacked *without altering the texture, the poise, or the straight line in phrasing.* It is quite easy to show this in black and white. In the florid tenor aria (Bach) quoted above, the long phrase may be broken at X and XX. If so, the semiquaver B♮ and the quaver E♮ respectively must be shortened in value. By halving the notes you *leave,* you can get a demisemiquaver and semiquaver of time in which to take breath and attack the new phrase with its full initial value. The loss of time-value in these is neither felt nor noticed, whereas the same taken from the subsequent semiquaver E♮ and D respectively would detract from the rhythm to an extent which would be painful.

In the fifth bar of the *bel canto* "O rest in the Lord,"

de - sires: O rest in the Lord,

the contralto will probably want to breathe after the word "desires" (she will probably have breathed three times already, but no matter), and it is essential that she should sing the passage in time because in this bar the flute plays with her in unison. She must, therefore, sing the B as a semiquaver, breathe on a semiquaver-rest, and start again with a full time-value on the A immediately following. She should, however, bear in mind that though the note A musically, and the word "O" dramatically, require that full time-value, their rhythmic importance is inferior to that of the G immediately following on the secondary accent, or third beat, of the four-beat bar. All singers —not merely contraltos—have some queer constitutional objection to relinquish the note they are on, only exceeded by the objection to tackle the new one; like other human beings, they dislike going to bed and hate getting up.

Let us now take the opening of a purely rhythmical song like Alessandro Scarlatti's "Già il Sole dal Gange."

Allegro giusto.

Già il so - le dal Gan-ge, già il so - le dal Gan-ge più

chia - ro, più chia - ro   sfa - vil - la, più chia - ro   sfa -

vil - la, più chia-ro, più chia-ro sfa - vil - la.

It will be seen that no breathing time of any sort is
provided for us by the composer, and yet the march,
or forward rush, of the rhythm is vital to the song.
If the singer is properly developed physically, and has
proper control of his breathing, he will breathe only
once, at X, making the first E♭ into a quaver (or even a
semiquaver), snatching a breath in the quaver or three
semiquavers of time which he has stolen, and picking
up the high E♭ following (without losing texture,
poise or the straight line) will give it its full value
and carry the phrase triumphantly to the end. Should
he not be man enough for this, he can fit in a breath
at the two points XX as well as at X, employing
the same note-poaching process. This is quite legitimate, and vastly superior to a gasp and scramble after
the more ambitious quarry, but it undeniably halves
the wings of the song — the heron somehow shrinks to
a curlew.

This brings us to the great subject of *size* — the
singing of songs in large. There is no sort of question
that the bigger the phrase, the bigger the song; and the
bigger the handling of the song, the greater the license
granted to the singer.

So far we have dealt with a main rule and its rigid
observance; with "you must not" rather than with

"you may." When "you must not" has been so
absorbed into the very system as to be forgotten, then
"you may" turns up and life begins in earnest.
Long phrasing—that is, long phrasing not only achieved
but assimilated and revelled in — is the essence of big
singing. Small phrasing narrows the range of vision;
the trammels of its physical limitations shackle enterprise and strangle individuality; hidebound by routine,
fearful of danger, its only vices peccadilloes, its only
virtues conventions, it lives and dies in suburban
orthodoxy. The long phraser sets no bounds to his
horizon; the *Wanderlust* is in his bones, and the whole
world lies behind or before him; hard work has
hardened his muscles and much walking has lengthened
his stride; loose-limbed, bright-eyed, self-reliant and
keyed by experience to emergency he conquers the
wild spaces of the earth, and Song, in gratitude to
the pioneer, presents him with the freedom of her
city.

Long phrasing is a matter of courage, pure and
simple. What physique and stamina are to the
pioneer, lung power and breath control are to the
singer; splendid in themselves, without courage they
are useless. The singer sees the big phrase ahead of
him, makes up his mind his lungs will not carry him
over it, funks it and halves it, and with it all that it
stands for. But long phrasing does not require
abnormal breath. In many cases abnormal inspiration
is a positive hindrance to phrasing, from the extra
muscular exertion required to control its expiration.
Long phrasing is simply a matter of will-power. The
singer can prove this for himself, and if he has
mastered his technique can master the long phrase in

F

a week. The greater his confidence the quicker the
mastery, and the quicker the mastery the greater his
confidence; it moves in a virtuous circle. Let him
take a broad beautiful melody such as Korbay's "Far
and high the cranes give cry,"

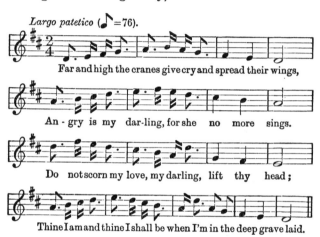

Largo patetico (♩=76).

Far and high the cranes give cry and spread their wings,

An - gry is my dar-ling, for she no more sings.

Do not scorn my love, my darling, lift thy head;

Thine I am and thine I shall be when I'm in the deep grave laid.

and make up his mind to sing each of the four phrases
in one breath; each must be sung *legato* and with the
crescendo and *diminuendo* carried in the rise and fall of
the music. He will probably begin by taking a huge
breath and holding on to it like grim death; three-
quarters of the way through it will come out with a
rush and blow the phrase to pieces. He will next
manage it somehow at a *tempo* rather quicker than
that marked, and without any great musical expression;
and in the course of a day or two that "somehow"
will have become "comfortably" and the *tempo* will
have broadened out. From then on the way is smooth
—"comfortably" turns to "easily" and "easily" to

"naturally" — and before he knows where he is, he is revelling in the beauty of the melody and his own power to sing it in large. Such a royal progress could never have been physical; no man could double his lung capacity in a week. It was "dogged did it."

Let him then move on to some big passage where not only is the singing of the phrase "in one" demanded by the music, but where actual illustration or appropriate colour within the bounds of the phrase is demanded by the sense of the words. The following passage from Stanford's "Fairy Lough" (An Irish Idyll) will do admirably:

On the fair - y lough a - sleep.

Here the singer must visualise the picture, and feel the eerie atmosphere of the little black lake which "lies so high among the heather." The wind never *blows* up there — the fairies do not like it; so the little waves move very slow, and in gentle little swells, and on them float the sea-gulls — *asleep*. The sea-gulls look as though they belonged there, as though they had once sailed in there from the sea years ago, and had settled on its bosom and stayed there ever since. Round and round the little green island they float and float, and as the little breeze dies down the little waves get slower and slower, and the sea-gulls fall faster and faster *asleep*. The two words "float" and "asleep" give us the atmosphere of the phrase. The little wave heaves gently up and down, and on it must *float* the voice getting slower and slower as it dies down with the breath of the breeze. The singer's first impulse is to take breath after "together" so as to give himself plenty of reserve to finish the phrase in large; but if he breathes there, the sea-gulls will not be "all together" any more — some of them will have woken up and moved away. The singer with his voice must tell you that they float round and round the little island all together all the time. To do this, not only must he sing the whole passage "in one," but — to give the feeling of peace — he must spread out the phrase from the word "float" and get gradually slower and slower down to the word "lough," and then — to show the utter absence of hurry and the human element — he should make a slight pause (almost imperceptible and without breathing) before "asleep."

This is of course no child's play, for merely to sing the notes *pp* and slow them down is not enough; the passage has got to have the natural *crescendo* and *diminuendo* of the rise and fall of the musical phrase on "round the one green island," and the subsequent accent on the word "fairy," and yet must sound sleepy and very far away "so high among the heather." He will feel at first inclined to give it up; he will think that the size of the effect is too small for the size of the effort; then he will feel ashamed, and, remembering "Far and high the cranes give cry," he will set his teeth and brace himself to it; then he will feel that gradually the necessity for that stiff jaw is fading away, that mind is getting the better of matter, that instead of hurrying over the hateful passage he is actually reluctant to leave it for its very beauty and the joy of painting it with his voice. As he sings, suddenly it comes to him like a flash that *bel canto* has lost its terrors; that the old herculean tasks are child's play; that his wings have grown and he is ready to fly, and somewhere down in the depths of his inner consciousness, as in a glass darkly, he will see struggling up a little thing called "style." Where has it been till now? Was it too, like the sea-gulls, asleep? Or was it there all the time, but, like Peter Pan, has it never grown up? Or is it not there at all but only reflected in the glass? One thing is certain: if it was born with him and grew up with him inch for inch of his stature, he would have felt all this without the telling.

Somehow all seems changed. He stands on the threshold of a new world. He looks for his old enemies "Keep off the grass!" and "Trespassers will be prosecuted!" and lo! they have vanished. Like the prince of fairy-tale he has found the key, the gates have swung back, and he has walked into the enchanted garden:

"He breathes the breath of morning pure and sweet
And his eyes love the high eternal snows."

He has grown suddenly from boy to man. His values have changed; he has forgotten the singing in the joy of song, and song has opened wide to him her golden gates.

The power to phrase in large exalts phrasing to a higher plane. The knowledge of its possession enables the singer to think in large, and therein automatically phrasing becomes interpretation — the means becomes the end.

The thinker in large — the master of style — is independent of rules. Rules have been so absorbed into his inmost being that the very thought of doing hurt to a song by their transgression offends his sense of fair play. His sense of touch keeps his balance true, and his sense of values regulates his proportions. But doubling the size of the phrase has worked some

miracle with the detail. In the small phrase (and the
small song) detail was everything; attention was concentrated on it for want of something bigger; singer
and audience took each little sentence in turn, put on
their spectacles, and hum-ed and ha-ed over the orthog-
raphy. When the whole was finished they found
that, in their anxiety to spell the words right, they
had forgotten all about the story.

Big phrasing and balance have changed all this.
Voice effects and petty illustrations shrink into their
shells at the sight of the big song. The succession
of *fortes* and *pianos* and *accelerandos* and *rallentandos*
which seemed to paint the scene so brilliantly just now,
have dwindled somehow to a thing of shreds and patches.
You can tell a master of style by his cadences alone.
There is no greater test of style than the finish of a song.
All singers have a tendency to sentimentalise their
cadences, to spin out the closing scene in an orgy of
rallentando. Not one song in a hundred, either in
text or in music, justifies, far less demands, such dis-
proportionate handling; yet "twice as slow" is a mild
estimate of the latter-end degeneration of the British
ballad. Like the long duck-roll to the indifferent
sailor, its elongated postponement precipitates the
inevitable nausea. It is the over-swing of the pendulum, the loss of balance, which does it. The master
of style, blessed with the overmastering sense of that
balance (prepared instinctively to tip up the scales if
he transgress), has fairly and squarely earned the right
to a free hand in interpretation. Saturated with
rhythm as he is, he may

(1) *Spread out, or narrow in, any phrase, or any
part of any phrase, anywhere, to any extent, at any time*

he likes. He is the sole judge. He knows that thereby he will not only heighten the emotional expression, but actually enhance the charm of the rhythm. Take, for example, the following passage from "Feldeinsamkeit":

und sen-de lan-ge mei-nen Blick nach o-ben, nach o - ben.

If the whole phrase be sung in one up to X (the end of the first "nach oben"), then the subsequent "nach oben" can be, and should be, lengthened out with a distinct *ritardando* and *diminuendo* in order to give the feeling of laziness and absence of worry. Both words and music demand such a broadening or "lazening" of the phrase; the voice fades off into a sleepy whisper, while the accompanist plays as though his fingers could not keep their eyes open any longer! The original rhythm, after its rest, starts then afresh with an even greater charm than it had before. Let, however, a pause be made at XX (as is practically invariable) for breathing purposes. The familiar gulp destroys the illusion. The sleepy eye lights up with a horror of rain, or at the thought that a pipe is not much good when the matches have been left behind. After such a perceptible breaking of the *sense* of the phrase, the subsequent *ritardando* will mean not laziness but boredom, and will convey that impression to the audience.

Again, in "Die Mainacht," also by Brahms, if the main sentences such as

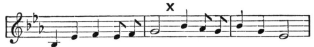

Wann der   sil - ber-ne Mond durch die Ge-straü-che blinkt,

und sein schlummerndes Licht   ü - ber den Ra - sen streut,

be sung throughout in one (*i.e. without* a breath at X),
not only can the singer give what elasticity he likes to
the phrase in large, and therefore to the song in large,
but he can give the proper grammatical values to the
text and, by the very size of the phrasing, covering up
the unquestionably false values given by the composer
to the unimportant words "wann" ("when"), "und"
("and"), "durch" ("through"), "über" ("over"), etc.,
he can focus attention on the pure beauty of the melody.

(2) *He may pause on, and lengthen out, any individual
note or notes when and where he pleases.* He may give
them an exaggerated value; but his sense of balance
tells him how far he can go. (Half a beat too much
will do it — if his phrasing were small he would not
dare — for half a beat is a lifetime to a little phrase.)

O the fun of that lengthened note, the holding of it
to the danger-point, the balancing on the edge and the
defying of distracted rhythm! And supposing there
are two such notes one following the other! or yet
another verse with two more such notes following
the first two! Poor rhythm will never take those
children to the sea-side again — her nerves cannot
stand it. And yet she is enjoying the excitement
herself more than any of them. Here we are in "The
Roadside Fire," by Vaughan Williams:

Where white flows the ri‑ver   and bright blows the broom.

Here the pause on "flows" is answered by a rather longer one on "blows." It enjoyed the excitement of standing on that slippery rock so much the first time that it means to stay even longer the second time, and to get nearer to the edge. These two are followed in the next verse

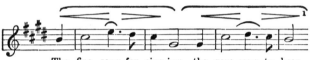

The fine song for sing‑ing, the rare song to hear.

by "song" twice, both "songs" longer than the "flows" and "blows" respectively, and the second "song" longer than the first; word balancing word, phrase balancing phrase, verse balancing verse, and one and all *pushing on* — even by the very holding back — damming the stream for a moment merely to make the head of water greater and rush it on to the inevitable end. The composer did not write these pauses; he knew better. He had no wish to see them like toy-balloons at the end of unlimited strings. He left them to the interpreter, as the wise man does.

(N.B. — *These effects must be rehearsed.*)

(3) *He may pause on a rest as long as he likes.* The word "rest" here is used figuratively rather than technically. It applies to the *fermata* on the final note

---

[1] The sign ⌒ is used throughout this book to represent an elastic pause or *tenuto.*

of a phrase, immediately before a rest *either marked or implied,* such as

im - mer   wo?   Im

Gei - ster-hauch tönt's   mir   zu - rück.

from Schubert's "Der Wanderer," or to the pause interpolated for dramatic purposes by the singer where none is marked, such as

der jun - ge Leib;   jetzt kenn' ich dich.

from Schumann's "Waldesgespräch"; or to actual rests, such as

from Schubert's "Der Tod und das Mädchen," in which, though the bars immediately preceding the rest have been played not sung, the initiative as to resumption is in the hands of the singer. Such elastic uses of the pause or rest are essentially dramatic and incredibly dangerous both to singer and song. By their very nature they seem to be in direct contradiction to all that this chapter has preached; but the master of style has his finger on the pulse of rhythm. He knows that such pauses, properly handled, enhance dramatic effect, arrest attention and stimulate magnetism. Pauses hold up a song; with them the singer holds up the listener as though he put a pistol to his head. The listener holds his breath; the singer's magnetic sense knows it. Let that sense play him false, let him hold that throbbing pause one fraction of a heart-beat too long and the listener sighs and looks away, the thread snaps, and song and singer come tumbling down together. This use of the pause is psychological; it is not to be played with by beginners.

Here we have the three great privileges, the three
great weapons of style — *elasticity of phrasing, prolongation of note-values, and the ad libitum handling of the
rest or pause.* Round them are grouped, or in them
are included, all the other helps to suggestion or
dramatic effect; tone-colour, word-illustration, climax
and anti-climax, variation of pace *in* the song (nothing
to do with *tempo of* the song), and all the other
variants of the singer's intellectual *rubato.* Some of
these will be referred to again later, but all belong to
the main three, and these three all belong to style —
to the song in large; the smaller the song, the less room
there is for them. We seem to have seen something
like them — caricatures — before in the *little* song.
They were called over-elaboration, cheap effect, and
self-consciousness.

Even the most patient reader must by now be
beginning to grow weary of the "big" and "little"
song; but the difference between them and the characteristics of each have been emphasised, or "rubbed in,"
for a very definite purpose. In interpretation, intellectually and physically, in style and in rhythm, the
little song and the big are as far removed from one
another as the poles.

This does not, of course, refer to the *little* song,
diminutive in stature but perfect in form — the song *en
miniature.* There are hundreds such written by all
the great masters, built on tiny lines to house a tiny
being. They are the smiles and frowns of sentiment,
the marionettes of drama. The singer may have to
put on his spectacles to see them; but he must handle
them just as lovingly as the big ones. Nay more, the
smaller the size, the closer the scrutiny and the greater

the care necessary. "Die Rose, die Lilie," from Schumann's "Dichterliebe," is more dangerous than the "Erlking," for it is over in a few seconds and the audience *has no time to forget*. For that very reason the diminutive song has not only to be treated physically and intellectually on the same lines as its big brother, but with even greater solicitude. Rhythm is the backbone of such songs. There is no room in them for dramatic effects or big contrasts. Their interpretation is almost invariably a matter of tone-colour assimilated to rhythm, and by rhythm they stand or fall.

With these may be included the large class of *"parlato"* songs in which the words are everything and the music but a delightful vehicle to carry them. Comic opera is full of them. The Gilbert and Sullivan operas teem with them, such as "My name is John Wellington Wells" from *The Sorcerer*, "The Flowers that bloom in the Spring" from *The Mikado*, and hosts of others. (Folk-songs, too, but these will be dealt with later.) They are nearly all quick in *tempo*, and all, without exception, are dependent upon strictness of rhythm for their musical effect. In comic opera they have, it is true, stage-setting and gesture as helps to that effect, but these are simply accessories; there are no big voice dynamics, no dramatic moments: they depend upon rhythm, pure and simple. It will be remembered that throughout this chapter the point has been emphasised that the spreading out and slowing down of certain phrases, so far from stopping the rhythm, actually enhances that rhythm on its resumption. How cunningly the composer of *The Mikado* handled his rhythm, and how surely he felt the pulse of the public, the very song quoted above proves. The man who has hummed "The Flowers that bloom in the Spring" (about 999 per 1000 of the Anglo-Saxon race) will remember the passage

and that's what I mean when I say or I sing, "O

bother the flowers that bloom in the spring, Tra la la la la, etc.

where the *tempo*, as befits an explanation, suddenly drops from fast to slow. If he remembers that, he will remember the gulp of delight with which he picked up and rattled off the old quick *tempo* the moment the particular line was over. It has never occurred to him to analyse his sensations. The composer knew him and prescribed for him. He knew that by holding those bars back, he held the breath and stopped the heart-beat, as it were, of the hearer, so that rhythm when it leapt into his blood once more pulsed, glowed and tingled to his very finger-tips. Again, at the end of the verse

Tra la la la la, Tra la la la la,

Tra la la la la la !

he holds up the rhythm, and again makes an over-powering effect by resuming it *a tempo*. Both are variants of the same process.

If the teaching of this chapter be true, and if the lesson of the above example be sound, we may count upon two things as applicable to the interpretation of almost every song in existence. (1) That practically every song has at least one point, if not more than one, in which it pays to spread out the phrase and "hold up" the rhythm, and (2) that it is not that "holding up" which pays in itself, but *the resumption of the "tempo primo."*

[The exceptions are those songs in which either the pictorial sense or the forward drive is more telling than any extraneous effect and therefore demands the absolute integrity of the rhythm; where, for instance, the monotony of the figure gives a particular atmospheric effect which a *rubato* would spoil (Luard Selby's "A Widow Bird"); or where its inherent strength is sufficient unto itself (Hugo Wolf's "Der Rattenfänger"); or both (Schubert's "Das Wandern").]

(1) The modern art-song is so fashioned that the singer need not bother about it; it is all done for him and probably indicated in the actual directions. But in the whole *a tempo* school, where the music and its rhythm are the essence of the song, and the *a tempo* dominates the *rubato* of interpretation, there is always some moment or moments where the song and the audience should be rested, and the rhythm either be given time to recruit, or allowed to "ease up" at the end of its race. There are numbers of ways of doing it, and the example quoted above from *The Mikado* is an excellent individual instance of the

general means; but the commonest method, especially in strophic songs, is by the *rallentando*, often accompanied by an individual *fermata*, at the end of the verse. In some strophic songs (though not as a rule in folk-songs) this is done in more than one (perhaps in every) verse; but the usual way in strophic rhythmical songs is to keep it till the last, and hardly anticipate it (if at all) before the end. In this connection let the singer remember what was said above — "you can tell a master of style by his cadences alone" — and not sentimentalise his "ease up" into a crawl home.

(2) To the finish of the song, the second point only slightly applies — and then chiefly to the concluding symphony; but to the "hold ups" in the course of it, it is vital. It is the resumption of the original *tempo* which is the fascination, the accumulated deliciousness of the "paid for!" after the more or less prolonged "on trust!" The singer must have the sense of deferred *tempo primo* ever in his mind as the key to rhythm and *rubato*. He has been "held up" himself often enough in the ball-room to know the joy of that first swish of the waltz when he gets back to it.

There are two ways of resuming the *tempo primo:* either, as in the above example, by starting it immediately after the prolonged note or phrase — generally a case of sheer high spirits; or, by a process of gradually working up the speed. How this should be done is a matter for the individual interpreter, and dependent on his sense of balance and proportion. (Singers should be on their guard, in this connection, when singing Schumann's songs. Schumann for some unaccountable reason, when he wrote a *ritardando*,

G

forgot as often as not to write the subsequent *a tempo*. The interpreter has, therefore, no definite directions to go by : he must simply use his discretion as to when and how to resume it, bearing in mind that the sooner he resumes it the less he endangers the march of the song.)

There are two types of song which seem to be individually either independent of this big main Rule I., or so written as to render its application physically impossible. The first of these is the song written in short sentences, generally in the form of question and answer, where the march of the rhythm is apparently subordinated by the composer to the exigencies of the dramatic situation. "The Broken Song" (An Irish Idyll), by C. V. Stanford, is an excellent example.

"Where am I from?" From the green hills of Er-in.

"Have I no song then?" My songs are all sung.

Here not only have we the series of short detached sections, but the composer has marked a pause, or *fermata*, at the end of each question. It is necessary to visualise the scene to get the dramatic significance of that pause. One can see the man gazing into the fire and thinking of old times, of the *"tempo felice nella miseria."* He repeats each question half to himself, thinks for a moment, nods his head once or twice, and then answers. The pause gives him *time to think*. In the tragedy of that question and answer the listener forgets the call of his rhythm, or would forget if he were asked to. But the composer has seen to it that the strain shall not be too great. Half-way through come a few bars of pure rhythmical beauty from the words "When she'd come laughin' " to "'twas the break o' day" like a smile through the tears, and then the tragedy closes round again up to the very end. In this compensation for, or suggestion of, forward rhythmical movement, the singer must help. His phrase may be arrested by the pause, but within its limits it must have the throb of rhythm. The song is written in quasi-recitative, and in recitative, as will be shown later, the rhythmic lilt of the phrase within its own compass is essential to its interpretation. The general effect on the listener of a song such as this is psychological. Its emotional power seems to overwhelm and absorb all the ordinary physical or even nervous effects; and what a chance for the tone-colourist! Question after question, answer after answer, running in tears and smiles through half the gamut of the emotions!

The second is what might be called the *moto perpetuo* song; that is, the song in which (quite rightly) no apparent possible breathing-time is marked either at all in the whole song (this does not of course apply to songs where no taking of breath is necessary, and which are therefore sung all "in one," such as "Quick! we have but a second"), or in which passages occur of such length that their singing in one is an impossibility, and of such speed that the first rule of phrasing (*vide supra*, p. 62) cannot physically be worked. An admirable example of the first of these is the beautiful pastoral bass aria from Bach's choral cantata, *Du Hirte Israel*:

Be - glück - te Heer - de Je - su Schafe, be -

glück - te Heer - de Je - su Scha - fe, die

Welt ist euch ein Him-melreich, ein Him - melreich.

Here we have a long aria lasting with the repeat (in
the vocal part alone) for fifty-four bars; without one

single rest, even of a semiquaver, in the voice-part
from beginning to end — a frank *moto perpetuo*. It
looks impossible to sing. It is not only possible but a
thing of sheer delight to the singer if he will but hold
fast by that first rule of phrasing referred to earlier.
He can breathe practically where he likes, by taking a
semiquaver value from the note he is leaving at such
points as those marked above. Owing to the deliberate
tempo, a semiquaver rest gives him ample time to
breathe comfortably on each occasion. He must never
at any one of these points pause a fraction of a second
too long on his rest, or not only will the beautiful
pastoral feeling of the song disappear, but the delicious
figure of the subject — moving in opposite directions in
the accompaniment and the voice — will be ruined.

A very good instance of the other is Ernest Walker's
"Corinna's going a maying." Here we have such
passages as

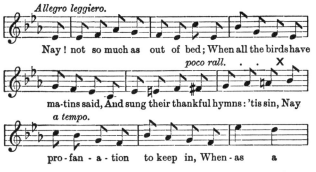

which, in spite of the rattling *tempo*, are impossible (with
their context) to sing in one breath, and in which, on
account of the same, it is impossible to breathe without
stopping the march. Where such passages occur, they

have to be specially phrased so as to bring breathing pauses into possible, or rather plausible, places in such a way — generally by a slight *rallentando* — as to convey the impression that the song was written that way and that there could not be any other way. But the *rallentando* must be slight and gradual and made to fit the sense of the words. Thus in the passage quoted above, the words *'tis sin* give the singer an admirable opportunity to pause in order to emphasise the point of the words, and at the same time to breathe comfortably. The usual place would be after *matins said*, before the word *and*, but this would involve a gasp and a scramble after the relentless (as will be shown immediately) accompanist, owing to the exigencies of the march of the song. It is thus possible to avoid the obvious, keep up the lilt, do it comfortably, and at the same time be apparently emphasising, and almost elaborating, a particular histrionic point. It is, in short, a species of justifiable pious fraud on the public. (In such songs as these rehearsal is of paramount importance.)

Accompaniment.

In the strict observance of this Main Rule I., the singer's greatest friend is the accompanist. There are still some people who say of the accompanist, "He followed the singer beautifully." Heaven help the singer if he did! If the singer knows that the accompanist will follow him, he will count upon it; the struggle with Nature will be too strong for his will-power and, fight as he may, he will find himself stopping the song to breathe. It has nothing to do with reading or interpretation; it is purely physical. If the accompanist is to do his duty and play fair to his colleague, he must *never wait for him* except as a definite part of interpretation agreed on and rehearsed by both. Upon that Spartan law and merciless help the singer relies for safety; knowing as he does that, if his breath fails, he will be left behind, he does not allow his breath to fail. The false relations of accompanist and singer are one of the evil legacies of the much-maligned Victorian period. Not so very long ago the accompanist in this country counted for nothing in interpretation. A dozen songs and a few arias carried the singer everywhere; the public demanded no more, and the accompanist knew them by heart. Voice and vocal effects, and the personality of the individual, were all that mattered; the individual sang, made his effects, stopped the song where he liked, picked it up where he liked, and the accompanist "followed." We have changed all that. The modern song is not a voice-part furnished with an accompaniment, but a work of art woven out of the two. Who would dare to say that the "accompaniment" of the "Erlkönig" or "Auf dem Wasser zu singen" did not rank in the one case equal with, and in the other higher than, the voice-part? What about Cornelius's "Ein Ton"? The voice-part, colour it as you will, would be a sorry thing alone. When would the contralto arrive at her destination in the "Sapphische Ode" if her colleague at the piano held back his syncopations to enable her to breathe the contralto number of times in the phrase? How many songs give the whole illustration while the voice simply gives the atmosphere; or the atmosphere while the voice gives the illustration? We have progressed since the days of the Alberti bass.

The accompanist who does not realise his responsibilities, who, from an hereditary sense of diffidence, refuses to assert himself and his individuality, is no good either to the singer or the interpretation. The very anxiety not to intrude atrophies the power to support. Such diffidence is supererogatory. He is invariably far the better musician of the two, and consequently to be trusted. Every singer who is worth his salt knows it, and only the fool dares to resent it. To the trained and level-headed interpreter the "following" accompanist is a drag on the wheel of his interpretation. The effort of carrying him along wears the singer out physically and mentally, for he has the double enemy to fight — his own physical weakness and the other's moral modesty. There is one other form of dangerous accompanist, namely the "pianistic" performer. This has no reference to his powers as a pianist — the finer pianist he is the better — but to his unconscious inclination to play his interludes, the bars which he has to himself in the course of a song, "pianistically" with the pianist's "Chopin" *rubato;* the danger being that such *virtuoso* passages, though admirable in themselves, upset the balance of the song mutually agreed on and rehearsed by both performers, and retard its march. For this and for every other reason, singer and accompanist should be old friends. What they mean to one another they hardly know. The full indebtedness of the singer to his accompanist is never known to any but the singer himself. He alone knows how often when he has felt slack and inclined to drag, when he has made a mistake, when his memory has failed him, when he has forgotten a word or begun too soon or too late, the other has pulled him together, got him back into his stride and held him there. He has started, maybe, without magnetism and without atmosphere; the other has coaxed them into his brain by the sympathy of his friendship and the magic of his touch. Let him change his *tempo*, or, as sometimes happens under certain subtle magnetic conditions, his whole reading at the last moment; his friend is with him in the new *tempo* or the new mood, his hand upon his pulse, doctoring him, stimulating him and cheering him on. The knowledge that he is there, safe as a rock — the stronger his individuality the better for his colleague — gives the singer powers he would never have had without him. They are partners in the same honourable firm, with equal shares; the longer and more thorough the association, the better for them both and for their art. For this reason the serious singer will see to it that, wherever it is humanly possible, his partner in private study shall be his colleague in public performance. The man he sings with in public should be the man he has worked with at home. Every note in the reading of every song is the property of each, and, if they have studied truly, has been tested by trial and adopted as the best by the deliberate judgment of both. That is true rehearsal, and on true rehearsal interpretation depends for its very existence.

The English concert system is responsible for a good many horrors, but the "local accompanist" has kept even the provincial "miscellaneous" concert at an unnecessarily low level for years. A concert party of, say, five singers and a violinist are engaged by a society in the provinces for a certain evening. The society has its "local accompanist" (probably engaged by the season). Each performer has the alternative either of bringing his own accompanist (which he cannot afford to do), or of singing, or playing, with the local man. He chooses the latter, and rehearsal is called for, say, 3 P.M. How in the name of common sense can any one man, however talented, *rehearse* adequately in an hour or two, far less perform later publicly, some twenty songs or solos? It is a laughable travesty of the word and an insult to good music. The singer, knowing the perfunctory conditions, puts in the programme the commonplaces of his *répertoire*, the banalities of the pot-boiler composer; and the poor public, thinking that because they are sung by the singer they are therefore hall-marked with authority, applauds them by night and buys them by day. Should the singer be too fastidious to stomach such stuff, he has to sacrifice some masterpiece, loved and nurtured at home, to be mangled in the *mêlée* — a horrible alternative. There are a few societies who even go so far as to make the singer's engagement conditional on his singing with the local accompanist. Such an obstinate loyalty is admirable from the point of view of parochial patriotism, but it is fatal to good work and most unfair on the man who wishes to do his duty by music. He has either to refuse the engagement or sing rubbish; he cannot afford the one or swallow the other, so he ends by trying to sing something good, worrying through a "scratch" rehearsal, and enduring an evening of shame and failure. The society notes the failure and does not analyse the cause or, if it does, ignores it, and the singer is not asked again. It must, however, be said that in most cases of failure it is not the accompanist who is to

blame. Accompanists vary as much in character and
temperament as do singers, and the "conscientious"
accompanist is no more good to singer or song than
the proverbial sick headache; but grant to him enterprise, character, temperament, drive, he has still to
reckon with the singer, and both with rehearsal.
Orchestral rehearsals must for economic reasons be
restricted in numbers — English orchestras read better,
learn more (and forget less) in one rehearsal than any
other in five — but there is no excuse for scamping
pianoforte rehearsals. The good song demands defer-
ential treatment. Without rehearsal interpretation is
a farce. Only the beginner in the profession or the
"florid" prima donna dispenses with it. The one doesn't
know that the music's the thing, and the other doesn't
care. The beginner has for so long concentrated his
attention, when studying, on his own part that he has
become an unconscious egotist. The prima donna counts
on the accompanist being as old a hand as herself and
her operatic *chevaux de bataille.* Her "Una voce" and
her "Jewel Song" have "galloped around the arena" so
often that she has some excuse. Well does the accompanist know the look of her music — punctuated with
pencil marks, riddled with red ink, scored with cuts and
repeats, and always to be transposed a tone up or down,
and generally without rehearsal! To the singer and
accompanist with whom this chapter has to deal, study
has been a joy. Rehearsal is a poor name for that
labour of love. The song has never been tossed aside
with a yawn, but, begun, continued and ended in
enthusiasm, has been stored away in the treasure-
house of both.

MAIN RULE II

SING MENTALLY THROUGH YOUR RESTS

RULE I. is physical in its application; Rule II. is essentially moral. Anybody can hear for himself, if he listens for it, whether the singer obeys Rule I.; no one can put his finger on the song and say, "So-and-so did or did not obey Rule II." Every now and again some listener, tuned to the singer magnetically, will feel its presence and give it a name; but the audience, in the main, passes it in the dark. None the less the audience is unconsciously moved by it, and the more cultivated that audience and the better the song and its singing, the more that subtle power stirs it. It is subjective in itself, yet it lifts the singing of a song out of the region of concert rooms into the fairyland of imagination.

Rule I. dealt largely with the phrase; Rule II. has to do with the song and the song's larger development, the song-cycle. It will be remembered that in order to sing the phrase truly, the singer (when studying phrasing) has to think of the end of the phrase first, then start it from the beginning and continue it to the end in a straight line, not pausing by the way — a species of non-stop run. *Unbroken continuity* is the essence of it. The song, as said earlier, is the phrase

in large; *unbroken continuity* is, therefore, the essence
of it, of the song as a whole, of the song in large.

Every song has certain beats or bars in which the
beginning or continuation of the interpretation is in
the hands of the accompanist, while the singer rests.
In such places the singer's *physical* voice ceases, but
his *mental* voice must go on. He must *sing mentally
through his rests*.

There are dozens of suitable analogies from the
world of motion, but one will do. The bicyclist, where
conditions render it advisable, ceases to work the pedals
and rides on a free-wheel; he ceases physically to
propel, but the machine moves forward none the less.
Once started, it moves on in a series of alternate
"pedallings" and "free-wheels," until its journey is
done and the rider dismounts. In the song, the
singer's voice "pedals" when he is actually singing,
and runs on the free-wheel while he is "resting"; the
song, like the bicycle, once started moves on until its
journey is done and the singer dismounts.

When does a singer begin to sing his song? *At the
first note played by the accompanist*. When does he
cease singing? *At the last beat of the last note of the
final symphony*. Every single note of that opening
and that close, and every bar of the intermediate
instrumental part, is not only his property but
his business. Every note he sings is the property and
business of his partner at the pianoforte. On that
interweaving depends the continuity of the fabric, and
on that continuity depends the song as a whole. The
singer must be standing and *thinking* at attention at
that first note, and remaining at attention till the last
chord has died away. He is singing all the time in

mind if not in body. The conscientious singer will often stand thus at attention for some appreciable time before beginning his song; it is to be hoped that his audience will not misjudge his attitude of mind. It is from no exaggerated sense of his own importance or of the deference due to him; he is simply waiting for that magnetic moment, known generally instinctively to his colleague and himself, when they and their audience are attuned one to another and ready to give and receive. To wait for such a moment of silence, to gather in the strings of attention, is the duty of the singer to song and audience alike. No one knows better than he that the song begun while the audience is "settling down" is lost, and with it more than likely all that follow it. Magnetism is a touchy person; if you pay her little court at the beginning, she may go off in a huff for the day. The "miscellaneous" concert with its long series of detached items is a poor school for Rule II. What inducement has the singer to give the rule a thought? In most cases the calibre of the song does not even deserve it. The singer entrusted with that song's unburdening upon a long-suffering public walks on to the platform with the music in his hand, nods to the accompanist to begin, finds his place while the opening bars are being played, worries through its commonplaces, shouts his high note at the end and retires mid applause, pleased and blushing (if he has a conscience), thankful that the particular species of pot-boiler has no concluding symphony to spoil the vocal effect. He cannot be blamed; it probably merited no better treatment. But if he, as he often does, essays something higher, does he aim at the higher level with it? The rush of the wind and the gallop of the horse in the "Erlking" do not find him ready; he is not shuddering with the father, shrinking with the child, or whispering with the "Erlking"; he is not even looking on; he is standing on the platform in the Queen's Hall or Albert Hall, or some other, hall, singing detached notes and phrases with great vocal ability and driving the individual points home. He has quite enough to do tackling those phrases without worrying himself about a piano part which does not directly concern him. What could Schumann have meant by those preposterous chords at the end of the "Two Grenadiers"? Just after the Marseillaise — finished $f\!f$ too! If the old soldier had had any gentlemanly feeling he would have broken a blood-vessel, or dropped dead some other way a few bars earlier, and given the singer a chance. If Schumann had only known the conditions, he would surely have altered it; in the meantime, better stick to the organ obbligato song as far safer.

The application of this Rule II. is largely a matter of temperament. The master of style has it by nature, and its application is unconscious; in fact, he could not be a master of style without it, as it is essential to the song in large. He is so absorbed in the song, so much part of it himself, and the song of him, that he cannot think of it in detachments any more than he can unscrew his arm to put on his gloves. Should the singer, however, not have it in him to begin with he should deliberately set to work to acquire and assimilate it. He will find that, though mental in itself, it carries with it certain physical conditions in its application. If he is standing at attention before the song begins and throughout its course, it is but a step from that attitude of mental attention to the condition of physical tautness. He will find — and if he does not, he must train himself to find — that, from the very fact of having *sung* the opening instrumental bars *to himself*, he is just as physically equipped (lungs charged, heart driving, brain visualising) to start his own part at the exact right rhythmical moment, as if every note of those opening bars had been sung by him aloud. So in the course of the song, after every bar of instrumental interlude, the march of the song finds him keyed and *poised* mentally and physically, ready and waiting to pick it up and carry it along like a runner in a relay race.

An excellent illustration of this interdependence of the mental and physical in continuity is Handel's "Angels ever bright and fair" (see p. 97).

This song is full of detached sentences, alternating between the voice and the instrumental part. These sentences or phrases have a peculiar clinging rhythmical drag, an inherent unwillingness to push on which seems to clog the wheels of movement. If the soprano here (as is generally the case) simply sings her own phrases and merely *listens* to the orchestral responses, she will unconsciously begin each of those phrases a fraction of a second too late; the act of pulling herself together in order to start afresh each time will have held her back, and by the time her first note of the fresh phrase sounds, the actual beat of the rhythm will have gone by. The detachment does it automatically; the result is invariably the same — the march of the song has been stopped. In the rare instances where the singer feels the call in her blood, where the pulse of rhythm and movement is in her to begin with, the effect is magical. The singing is lifted from performance to interpretation; the limp is gone, the song moves with the grace of a minuet, and the audience, carried along by its lilt, sighs at the end, it

fair, Take, oh take me,   take, oh take me to your
care,   take me
etc.

knows not why, and thanks the singer for a master-piece. That style of singer has no wandering eye or collapsible lungs. Her eyes are on the ultimate goal and her lungs are keen to push on thither. There is no loitering by the way for her. She is ever ready for that rhythmic beat — almost to the extent of

H

anticipating it — and to let it go by would hurt her to the quick. The conductor can be a great help if he will. He generally is, supposing he has not abandoned the effort in despair. If the singer plays her part, he may generally be trusted to play his. Of all slovenly faults, "waiting for the beat" is the most bedraggled.

Let us now apply this Rule II. to the higher type of aesthetic work and see how it affects it. "Der Leiermann," by Schubert, is, as said earlier, an essentially atmospheric song and correspondingly interesting as a study. The voice here is subjective. There are two main points of vision in the song — the hurdy-gurdy man and his hurdy-gurdy. The piano part is the hurdy-gurdy; but the singer is not the hurdy-gurdy man. If he is anyone, he is the man who watches the hurdy-gurdy man from afar. Now the hurdy-gurdy is objectively illustrated in the actual music; the man is not — he is simply spoken of. Hence, if the mentality of the singer is to absorb either one or the other, the hurdy-gurdy has the greater claim — and the hurdy-gurdy is in the accompaniment. Therefore the singer cannot ignore the accompaniment and concentrate his mind upon the quasi-narrative vocal part without committing a grave injustice. As a matter of fact the use of the voice throughout is purely atmospheric, and in its actual words merely represents the musings of the onlooker. The singer colours those words to represent the impression made by the scene on the onlooker's mind. But the speaker (the onlooker) did not receive his first impression from the sight of the old man. He has ears as well as eyes. Long before — eight bars before — he saw him, he heard his hurdy-gurdy. The

latter gave him his first impression; its drone fasci-
nated him; its wheezy old lilt made him search, to
try and mark it down. He has eight bars in which to
find it. Anywhere within those eight bars — let us
say at the sixth — he sees the old man. You can
actually feel him absorb the atmosphere of the scene,
the dreary hopeless picture of the old automaton-man
turning the handle of his old automaton-machine,
grinding out the same monotonous old tune, ever at
the same pace, the same to-day as yesterday, and
to-morrow and every day until both are worn out and
dead. To the singer who is a tone-colourist and who
has the sense of atmosphere, the hurdy-gurdy has the
prior claim. To him as a singer those first eight bars
come first, and he will *sing* them first, as part of his
picture, not physically, of course, but mentally. If
he does not, then assuredly the song will be as dreary
as its subject, though not precisely in the form he
meant it. Superficially the singer has nothing to do
with the singing of the song until he begins to sing;
but the man he then begins to sing of has been turning
the handle for eight bars already to the consciousness
of the narrator. If he does not realise the meaning of
the very first bar and its drone, he can never give to
others that sense of dreary monotony epitomised and
canonised by poet and composer.

The above shows the importance of mentally realising
the objective or directly illustrative instrumental subject. The same holds good when the parts are reversed.
There are songs in which the voice gives the illustration
and the accompaniment the atmosphere. Here the
singer, if he is to illustrate truly, must have mentally
absorbed the instrumental atmosphere from the be-

ginning in order to start in the right mood. Such an example is Charles Wood's "Ethiopia saluting the Colours" referred to earlier (p. 15).

It is not until the second verse that the hearer is actually informed that the speaker is a soldier and his regiment on the march: but the first bar has told it to him and the interpreter by implication. To make himself into that soldier the singer must have sung those opening bars in step with his regiment, and throughout the song — half a ballad, half a contemplation as it is — the tramp of marching feet is ever in his ears. But the interest and incident of the song are focussed on the figure of the old woman; the accompaniment is subjective. The song is the exact opposite of the "Leiermann"; but in both cases, from the first note to the last the song is sung and played by singer and accompanist together as a whole.

In Cornelius's "Ein Ton" would any singer contend that the musical interest is concentrated on his own five phrases sung on one note? Could he of himself assimilate and express the five colours of those five phrases without the stimulus of that simple and wonderful accompaniment? Is he prepared, when he has sung his last note, to call in his thoughts and let the instrumental end take care of itself? Or will he sing every note of that final symphony, and dwell on it and let it go reluctantly at last? If he will, he may be trusted — even with the "Dichterliebe."

To the superficial singer Rule II. will not appeal. There is nothing to show for it, nothing tangible as reward for his trouble. Its effect is far below the surface, far too deep to handle consciously. Yet it is there. Let him wait until his audience is at rest

and their attention concentrated on what he is going
to tell them, and let him begin then and not till then.
If he will but sing those opening instrumental bars to
himself, that attention will never leave him, for he
has also sung them to his audience. Let him, on the
other hand, experiment. Let him begin his "Leiermann" or his "Ethiopia" at random, with audience
and himself unprepared, and watch the result. He
will notice that when he sings his first note the
wandering eyes which have been criticising the cut
of his coat, or silently sympathising with his figure,
will drop of a sudden to the book of words to see
what it is all about; the skip from London to Bavaria
or Georgia will be too big a hiatus to tackle in the
time, and the song will be sung and over, dead and
buried, before the singer and his audience have ever
joined hands. The observance of the rule has its
immediate effect in application; its effect on the singer
is far deeper, for it is not only a pledge to himself of
sincerity — but it drives out self-consciousness with a
pitchfork.

Poise.

The singer who obeys this rule is spoken of above
(p. 96) as being "poised" mentally and physically
in the line of the march of his song. Every song
feels to the singer to be upon a certain level and to
move forward in its straight line upon that level. It
runs, as it were, upon telegraph wires; the poles may
be long or short, and at varying distances, but the
wires run in a straight line and at the same distance
from the top. This poise, or keeping of the level, is
a feature of style. It seems to have been taught or

appreciated by the old Italian *bel canto* masters, for
it is conspicuous in the work of the best exponents of
that school. It is partly physical but chiefly mental,
the physical application being probably consequent
upon the mental attitude. Like the main part of
Rule II. it is not visible on the surface, and the
audience though conscious of its charm does not give
it a name; but one singer recognises it in another and
takes off his hat to it. Songs have a mental as well
as a vocal *tessitura*. These have nothing actually to
do with one another, except that both indicate a region
in which the song should lie. The key to the mental
tessitura is generally one note, round which all the
other notes seem to group. Through the whole song
this one note seems to run (like the B in "Ein Ton"),
and on this note the level of the song is poised.
Thus in "Angels ever bright and fair" the mental
tessitura seems to lie round the fifth (C); in Mendelssohn's "Auf Flügeln des Gesanges" in G, the note
seems to be the third (B). In Korbay's "Mohàc's
Field" (in D) the note is unquestionably the tonic
D, which dominates and sways the whole song. In
"Der Leiermann" (in A minor) the note is again the
tonic A unmistakably. The monotony of the drone
on the tonic with the open fifth goes through the
whole song like a pedal point. The natural selection
of such "poise" notes must vary, no doubt, with the
singer. Like the master-phrase in atmospheric songs
he probably need not look for it; it will come of itself.
But if he can once get the insistent call of it into his
ears, it will have a direct physical result in the actual
singing. He will find that not only will the song
seem to be wound up and running like clockwork on

its wires, but that the sounds themselves will seem to come from one particular point in the actual sounding-board of his head, the point to which his "poise" note called them. That means control, concentration, economy, "push on," and all the other friends of technique and interpretation, and puts an end to the type of work known as "all over the shop."

Rests are as much a part of a song as notes; they are necessary to the singer physically, both as breathing-places and actual relief from vocal strain. But he must make up his mind to one thing — that from the moment the first note of his song is played there is not a rest of a fraction of a beat to his brain. Once started, he must go through with it to the bitter end. The singing of a song is a great responsibility and a great strain; on his ability to accept the one and stand the other depends his power to carry it through.

The song-cycle is but the song in large. From the first note of the first bar of the "Dichterliebe" to the last note of its wonderful final symphony, the singer knows no rest — he is singing mentally through them all, as well as physically through all his phrases. Thirty-five minutes of constant strain, singing, visualising, magnetising, driving, pushing on, and not one rest!

MAIN RULE III

SING AS YOU SPEAK

ALL the singer's gifts, all his perfection of technique, all his observance of rules go for little or nothing if his singing is not speech in song. For this he must have

(a) Purity of Diction.
(b) Sense of Prosody and Metre.
(c) Identity of Texture in the sound of the spoken and the sung word.

(a) PURITY OF DICTION

The Anglo-Saxon race is the most good-natured in the world. It has a peculiar horror of wounding the feelings of the public performer, and out of sheer good fellowship it puts the telescope to its blind eye rather than hurt. This happy relationship between performer and audience has made public singing a pleasant life for the professional wherever the English language is spoken. But like every other form of *laisser aller* it has worked for harm unconsciously. The "*basta! basta!*" of the Italians may be cruel in its immediate application, but it is cruel to be kind. It

104

is a protest against the lowering of certain standards long accepted by the people as compulsory. The singer who has not trained himself to them knows what to expect. He makes no complaint and generally tries to right himself; for he has had the inestimable benefit of directly realising where and why he falls short. The British singer (which may in greater or less degree be taken to mean the Anglo-Saxon) has suffered badly from the other extreme. For his faults of diction he is hardly to blame, seeing that his public has never remotely demanded a standard of him. That public has either never realised that it has a language of incomparable beauty of its own, or has from long acquiescence in bad habits learned to accept those bad habits as inevitably associated with the musical expression of words. It may have vaguely felt that its particular sweetheart, male or female, would have looked distrustfully at it had it spoken words of "lorve," but had it sung them instead, the sentiment would have been automatically exalted from vulgarity to poetry. In any case, if it has its suspicions, it keeps them to itself, and out of sheer good-nature suffers singers gladly. What is the poor singer to do? He has no one to keep him up to the mark. When he stops the song to breathe, he is rapped over the knuckles by the accompanist and takes care consequently not to do it again; but the fellow-countrymen of Shakespeare and Milton ask of him nothing, and, after all, he is not primarily a philanthropist. The old Italian school of singing was a pure joy of sound. In many cases they had not much to say, and the text of what they said was valueless; yet every note was a gem of beauty and every word a model of elocution.

But the rigid adoption of that school for the training of our singers was another evil of the Victorian period — not vocally, but in the matter of language. The Italian language is very limited in the number of its vowel sounds, though each vowel is purity itself; the English language contains almost every possible modification of every possible vowel. Italian consonants have a liquid incisiveness (to be paradoxical) which is almost impossible to transplant into the solid mouth of the Anglo-Saxon. To the Italian singer these vowels and consonants come by nature as the direct translation of his beautiful speech into beautiful song; the English singer, in his endeavour to assimilate them, has merely absorbed their limitations and missed their characteristics. He has remembered the "Voce, voce, voce" of Rossini, and in the desire to *sing,* he has forgotten to *speak.* The consequence is that English singing is dominated by the hybrid vowel, the compromise between the Italian and the Anglo-Saxon; and by the carried-on consonant, the solid English version of the liquid Italian original. The public, as said above, has heard them so long, and looked upon them as so sanctioned by authority, that it has never felt even remotely entitled to anything better. Here is a typical version of a few lines of a popular song — Barnby's "When the flowing tide comes in" — as sung by the contralto and applauded by the public. It is terrible to look at and more terrible to hear, but it is not a caricature.

"Mawther-a," he cry-eed-a, "gaw wortch-a tha ty-eed-a,
Arz it cawmeth-a arp-a too Lynn-a.
For-a fou-url-a or-a fayr-a oi weel-a be they-ra
When-a tha flaw-inga ty-eed-a cawms in-a."

What must the true Italian think of such stuff as this? or the other purist, the Frenchman? With no knowledge of our language he could do it better himself by sheer virtue of discrimination and sense of refinement. He does not know that that particular type of British contralto, by far the worst offender in every branch of singing — rhythm, technique, intelligence and diction alike — has only one object — to make her voice sound. To do this she has to struggle, it is true, with unaccommodating registers, and the telling portions of her voice are limited; but her proximity to the note on which she expects to make her most brilliant fog-horn effect can be gauged by the brightening of her eye. What is diction to her? The trombone-like blare (often accompanied by a double slur) in the penultimate bar amply atones for the jargon of the rest. It does not matter to her whether it was she or Brünnhilde who was "seated one day at the organ," so long as her "Amen" sounds "grand." There are, of course, many exceptions, and their power over audiences is correspondingly remarkable, but it must be confessed that for shortness of breath, incapacity of phrasing, slovenliness of diction and narrowness of perspective, this type of British contralto stands in a category by herself. The bass, alas! is a dull dog as a rule. He suffers greatly from manliness and cathedral traditions. Anthea would have been unfavourably impressed with the effeminacy of the man who preferred to "die" rather than to "doy" for her on a broad manly vowel; while "Aw-mairn" and "hawly" are as much the accepted cathedral-bass pronunciation of "Amen" and "holy" as "saw-url" of "soul." The fifth verse of the "Te Deum" in the average church service would be

laughable if it were not a travesty of great words —
"arnd crorse tar-r-rknaiss-a the peopurlla" may be
heard any day in the *Messiah*. The baritone has
much to be thankful for. Nature has allowed him
to be born without any inherent predilection for
long hair or butterfly ties, and has endowed him
with more actual gifts of tone-colour, broadminded-
ness and sense of words than his fellows; his super-
fluity of numbers too (99 out of 100 male voices
are baritone) has found him his level and trained
him, by competition, with the thoroughness of a
public school. *Mutatis mutandis* the same applies to
the mezzo-soprano. These two may be best trusted
to rise to the level of their responsibilities. Sopranos
and tenors sin more by omission than commission. In
diction they simply avoid difficulties, and consequently
their version of their own beautiful and varied language is more or less colourless. The higher registers
in which their voices are most effective do not lend
themselves so easily to vulgarity, and their mal-pro-
nunciations are therefore not so offensive to the sensitive ear of the purist. (At the Crystal Palace a few
years ago "aharrawarradedda" was once perpetrated
by a well-known tenor, with complete confidence and
entire success, as an equivalent for "rewarded"! An
eminent musician who was present noted it down
phonetically for the benefit of posterity.) They have
a tendency to shade, at the smallest provocation, all
vowels to the open quality, sacrificing thereby not only
the charm of variety but the fascination of such pure
deeply expressive closed vowels as the "ee" in "meet,"
the "oo" in "swoon" or the "u" in "pure." The
soprano cannot, of course, sing the pure "ee" on a

top A — it would sound like a slate-pencil — any more than the bass can sing the pure "a" in "part" on his top E without not only making it sound ridiculous but actually damaging his voice. The shading of the one towards "ah" and the other to "awe" is automatic, and, being natural, is accepted by the hearer as right. But the cult of the open vowel has developed into a religion — the religion of the praying-wheel, *of the line of least resistance;* while language is disestablished and disendowed. The open tone is superficially the easiest and most natural for the human voice. The average voice has probably been trained principally on the open "ah" or hybrid (though most useful) "aw," and any deviation therefrom to the closed and modified vowels is regarded as a nuisance, to be avoided or compromised. Thus "meet" becomes "mairt," as a compromise between "meet" and "mart"; "soul" becomes "sawl" either as a compromise between "soul" and "sarl," or as a direct avoidance of the "o" and adoption of the "aw" — and so on. Here are a few of the hybrid pronunciations of Anglo-Saxon singers. For reasons of space their numbers must be restricted here; their name in reality is legion.

man	= mahn.		best	= baist
swan	= sworn.		gay	= gair.
cat	= cart.		them	= thurm.
dog	= dawg.		neck	= nurk.
horse	= horr-se.		read	= raird.
(Like the "o" in horror.)			wheat	= whirt.
rack	= rark.		pin	= peen.
and	= arnd or ur-yend.		kiss	= keess.
men	= main.		his	= hez.
blest	= blairst.		if	= eef.
slate	= slairt.		wind	= wa-eend.

smoke = smawk.   rose = rawz.
slow = slaw.   moon = murn.
moan = morn.   juice = jawce.
coat = caught.   coot = coat.
close = claws.   Etc., etc.
snow = snaw.

The "ee" and "oo" compromises are sounds almost impossible to express on paper. The above are about as near as one can get. (These two close vowels, being particularly effective with basses and baritones, are treated by them with respect.) When any of them are followed by "l," an additional curl of the tongue, involving an added vowel, is given, making the pure vowel into a horrible diphthong, such as

all   = aw-url.   cool = coo-url.
holy   = haw-early.   self = sair-lurf.
soul   = saw-url.   (In musical comedy,
almost = aw-url-morst.   say-url-urf-a.)
dwell = dway-url.   full = foo-url.
tell   = tay-url.   Etc.[1]
creel = cree-url.

In this basses and baritones are by far the worst offenders. As no British ballad is complete without either a "love" or a "rose" or a "soul" in it, the purist who listens to it is assured of at least one black eye, generally followed up by severe punishment all round; if he is wise he will leave the ring early. The singer who makes "glow" into "glaw" and "stow" into "staw" might presumably be trusted with "glory" and "story," but for some reason — apparently from sheer cussedness — she sings them as "glow-ree" and "stow-ree"; dislike to certain values in the right place seems to superinduce them in the wrong. Such versions of "pleasure" and "treasure" as "play-joor-a" and "tray-joor-a" are to be heard any night in musical comedy; but this particular form of performance is so saturated with plague bacilli that only the British constitution could stand it.

Closely allied with the pure vowel sounds are the diphthongs. These, as everyone knows, are made up of two vowels, of which the first is related in importance to the second in the proportion of about nine to one. Diphthongs are of two kinds, actual and implied; actual, as in the "oi" of "rejoice"; and implied, as in the "i" of "mind." The treatment is the same for both. The first elementary rule of diphthong-singing is that the primary vowel should be given practically the whole time-value, and the secondary only so much as is inevitable in getting away from the first, either to the consonant following it, if there be one, or to the finish of the word. This secondary value should be so small that the hearer should never feel conscious of its demi-demi-semiquaver share having been taken from the note at all. Thus in "rejoice" the diphthong is composed of a primary rather bright "aw" followed by a secondary "ee." Should they be given anything like equal values the passage from "Rejoice greatly," in the *Messiah*, would not only sound dissyllabic instead of monosyllabic as follows,

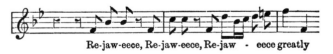

Re-jaw-eece, Re-jaw-eece, Re-jaw - eece greatly

the crotchet being actually converted into two quavers, but the illustrative meaning of the word through the call of the open first vowel would be destroyed and

vulgarised as well in the process. The "i" in "life" is
made up of "ah" and "ee"; that of "wait" of "ay"
and "ee," and all the others in the same way; all
must be treated alike. Yet it is possible to hear

La-eefe for ev - er -more.

(" Nazareth," Gounod) from a bass, and

way-eet   pa -tient-a-ly   for-a Him.

(*Elijah*, Mendelssohn) from a contralto, both volunteered
with complete and confident *bona fides* for "English as
she is sung." The word "wind" is given a value all
its own — probably a survival from the Italian methods
of our fathers. Not only is its vowel generally not
given the sound as in "swim" or "spin," but it is
actually converted into a diphthong, "wa-eend." (Its
occasional rhyme in poetry with such a word as "mind"
is purely a rhyme of eye not of ear.) The above
diphthongs are pre-consonantal; there are others which
come at the end of words and melt into thin air, such
as "toy" — in which the "o" sound (as in "lot") is
followed by "ee." Here the secondary vowel must be
given even less value than in the above words, if that
be humanly possible. It must be treated almost as a
parasite and flicked off with lightning speed. The
treatment of the diphthong immediately before the
final "r," as in "hour," will be dealt with presently.

(N.B. — In all the above vowel-illustrations, the "r"

is used simply as a guide and is not meant to be sounded.)

Aspirates speak for themselves; there is no temptation to handle them differently in song from speech. There is a subtle use of them which will be dealt with under "word-illustration" (p. 184).

The uses and abuses of consonants are hard to put in writing. To find the happy medium between incisiveness and demonstrativeness is the difficulty. The British singer, as a rule, is inclined to underdo his consonants, the German to overdo them. But the German is trained in a finer school. If Wagner had done nothing else, his achievements in musical elocution alone would have left his country for ever in his debt. Of all composers Wagner wrote the truest vocal music from the point of view of dramatic diction. The actual physical powers required to sing his operas are, no doubt, abnormal, but his works for purity and ease of declamation are never likely to be surpassed. The modern German school, with that before it, does not fail in distinction of utterance; it often errs on the other side and sacrifices beauty to declamatory strength. (There is a type of German singer known as "Konsonantensänger.") Those English singers who have had a German training can be trusted not to shirk their duty as elocutionists; some even acquire a certain sledge-hammer brutality of diction which is not far removed from sheer vulgarity. To find the happy medium is, as said above, the difficulty. The master of style feels it, like most things, in his blood.

Consonants are practically "embroidered" on the column of air of the sounded vowel. That column of air comes (or should come) out in a straight line which, when composed of the right vowels and consonants made into words, constitutes the straight line of the phrase. To the singer whose diction is in the right place, namely, the region of the tip of the tongue and the lips, the initial and intermediate (or mid-word) consonants present little difficulty; his sense of touch will give them their right proportions in the word in its relation to the sentence and its dramatic significance — provided, always, that his native English dislike of demonstrativeness does not make him too shy to do so. It is the final consonants which give most trouble. The ending-off of the word without illegally carrying over to the next, and the preservation of the straight line and "push on" of the phrase, are so opposed to one another in practice that they seem almost incompatible. The Italians have a liquid way of their own of doing it, which is so sanctioned by custom as to have become legal, and so refined in its handling as to be void of all offence. But it is a way which does not assimilate readily with the roast beef of old England (the Englishman's Italian "carry-over" is about on a level with his Irish brogue), and the results are generally such indigestible solecisms as "Seated-a one-a day at the organ-a, I was weary and-a ill-a at ease-a," or the Barnby song quoted earlier. (Excess of zeal has applied this even to mid-word consonants. The following delightful instance of *trop de zèle* has been heard in our concert rooms:

Allegro con brio.   Carissimi (?).

Vitt-it-tor-ia, vitt-it-tor-ia,   vitt-it - tor-ia, vitt-it-tor-ia.

where the over-anxiety for incisiveness has converted
Carissimi's original crotchet into two actual quavers !)
The crux is, of course, the straight line and forward
movement of the phrase. (The difficulty of reconciling
this with the natural halt of diction has so frightened
some singers that they have involuntarily turned those
first $\frac{4}{4}$ bars of the "Lost Chord" into the easier slow $\frac{6}{4}$,
from sheer inability to face the monotony of their slow
march.) The perfect union of the two depends primarily
on command of the first branch of technique — breath
control. Unless this is so firmly in hand that the
actual body of vowel sound of the phrase can move
forward in the straight line independently of consonantal interruptions, the reconciliation can never
take place. In singing the vowel is the predominant
partner — as the very derivation of the other implies —
and its work must never be stopped by petty inter-
ferences of consonants. The English singer can put
the consonant in the proper place if he will only sing
it as he speaks it — as he speaks it in English not
Italian. Consonants require no sledge-hammer to drive
them home; the music does that better than the
singer. All he has to remember is that the mumbled
jargon of the ordinary English conversation is not
speaking in its true sense, and that the speech which
he has to exalt into song must be as clear-cut and
clean as the music it adorns.

Finally there is the treatment of the letter r. There
is no question that the r, wherever it comes in the
word, ought by rights to be sounded. It was not
originally put there for ornament, and, as a matter
of fact, in Ireland and Scotland, where the best English
is spoken, it is treated as an essential and almost

invariably given its value. But having granted it
its rights we have to see how far we can reconcile
them with our own customs. If singing is to be
speech in song, we must not, from any spirit of rigid
purism, read into that speech effects which are foreign
to the accepted use of its language, and there can be
no doubt that for practical purposes many written *r*'s
have fallen into disuse in pronunciation. The rolled *r*
has been misused by both professional and amateur
from time immemorial; by the first as a cloak to
cover a multitude of sins, and by the second as a
passport to professionalism. There is no sort of
question as to its importance — the professional singer
who cannot pronounce his *r*'s is practically unknown
— but it can, and should be, honoured almost as much
in the breach as in the observance. It is a valuable
asset in songs where the effect aimed at is virile, incisive,
and strongly declamatory. The word *proud* seems to
demand a certain roll of the *r*; this comes here at
the beginning of the word and carries a certain strength
with it. In practically all words in which it comes at
the beginning, such as *friend, priest, fresh, cruel, crash,
cringe, branch, dread*, etc., it should be strongly sounded.
When it comes in the middle of a polysyllabic word it
should be sounded or not at discretion, in accordance
with the accepted spoken pronunciation. Thus in the
very word "disc*r*etion" it should certainly be sounded,
whereas in "inco*r*porate" the sounding of the first *r*
would be as gratuitous as that of the second is indispensable. Wherever it appears as the connecting
consonant between two vowels, as is this second *r*, its
sounding is imperative. (This applies to dissyllables,
not to words like "cared," "feared," etc., in which its

sounding is not invariable. These are, of course, in
reality dissyllables too, but have become monosyllabic
in colloquial use.) In a word like "prīmrŏse," both
r's would be sounded, though the first would be far
the stronger of the two; this is not so much a question
of comparative *r* values as a concession by the second
r to the lilt of the Trochee — *prīmrŏse*. In Trochees
where the *r* comes at the end of the short foot, as in
eāstĕrn, *hōmewărd*, etc., it may be slightly sounded
as a small compensation for the shortening of its
preceding vowel; whereas in Iambics such as *ăstērn*,
rĕwārd, etc., rolling it would be, if not actually vulgar,
at least uncalled for, and would detract from the pro-
sodic strength of the foot. It will be seen from the
above examples that, with the exception of the particular type of Trochee quoted above, the most strongly-
rolled *r* is the one which comes either *after* another
consonant or between two vowels. The rolled *r* *before*
another consonant is generally the reverse of pleasant;
the word "charm" loses all its charm if the *r* is rolled;
likewise in "hea*r*t" it is unthinkable, but it is sung
thus in public *ad nauseam*. *Storm, horse, snort* from
their very strength seem to require it, and yet they
are generally better without it. The rolling of the *r*
in "ho*r*se" seems irresistibly to force the vowel from
the *o* (as in "corn") to the *o* (as in "lot"). The subsequent transition from "ho-rrrse" to "ha-rrrse" is
short. The rolling of the *r* at the end of a word is,
with one exception, impossible, and the carrying of it
over to the next (though quite common) is as much a
solecism as the interpolated *r* of "Emma-r-Ann." The
exception is in the case of such words as *our, here,
poor*, etc., in which there is always a danger of making

the word dissyllabic or diphthongic. The slight roll at the end resolves the vowel without allowing it to drop from its level. Such words are, of course, theoretically diphthongic, but so mono-coloured as to be practically monosyllables. *Ow-ur, poo-ur, hee-ur* would be terrible. This also applies to such words as *near, clear, cheer*, etc. Here the purist frankly shies at speech. He never could, and never will, transplant into song such horrors as the *ny-ur, cly-ur, chy-ur* of conversation. He has compromised by treating them in the one case ("cheer") as a pure single vowel ("ee"), and in the other ("near," "clear") as a diphthong ("nee-ar") whose primary and secondary vowel values are left to the singer's taste and discretion — which, after all, govern and administrate the whole business in the end. It was no doubt the fear of such "conversational" effects that drove our forefathers to the other extreme, and made them write *Pow'r* for *Power*. Where meant to rhyme with *hour* (as no doubt has been done) this eliminating process would be inevitable, but *Power* is primarily a dissyllable, and the dangers of *Pow-wur* can be easily avoided by the singer, as the next portion (*b*) of this Rule will show.

One concession has to be made by speech to song. Where a vowel is held for a long time — especially if sung *forte* — on one note, it assumes an instrumental character; roundness and fulness of tone being here essential, a certain latitude of broad effect is allowed it, even at the expense of pure phonetics.

Max Müller in his *Lectures on the Science of Language*, makes the following trenchant remarks: "There is one class of phonetic dangers which take place in one and the same language, or in dialects of one family of speech, and which are neither more or less than the result of laziness. Every letter requires more or less muscular exertion. There is a manly, sharp, and definite articulation, and there is an effeminate, vague and indistinct utterance. The one requires a will, the other is a mere *laisser aller*. The principal cause of phonetic degeneracy in language is when people shrink from the effort of articulating each consonant and vowel; when they attempt to economise their breath and their muscular energy."

That is a criticism which though written of speech applies in its entirety to song. Why should the contralto give us this? ·

:"Ly-eek arz the darmarsk rawse you see

·   ·   ·   ·   ·   ·

Or ly-eek arn ow-ur or ly-eek a spahn
Or ly-eek the seengeeng orv a sworn
E'en sorch is mahn — who leeves by braith
Is hee-yur nee-yur dairth — mahn's ly-eefe ees dawn."

She would not *speak* it thus in the bosom of her family (except, possibly, to illustrate the Bogey-man for her children); why should she *sing* it thus in public? It is sheer laziness — slovenliness, if you will. She and her colleagues have avoided the "effort of articulating" for so long that they have forgotten how to do it. For years they have followed the cult of the hybrid vowel and the shirked consonant — the line of least resistance — and the line of least resistance is as far removed from true singing as the North Pole from the South. Singing is one long physical effort and mental strain, and the man who pretends to sing without them is an impostor. He is not wholly to be blamed. The public has encouraged such imposture

by its own inertia. It has just sat there, heard,
yawned maybe, and turned over the page to the next,
but shown no active disapproval. We would not have
it otherwise if we could. That inborn courtesy, that
sympathy for the other fellow, that spirit of fair play
is too great a national asset to lose. The singer must
work out his salvation for himself, remembering that
if one plays fair, so must the other. He has the finest
language in the world, the language of his Bible and
his forebear Shakespeare for his heritage, and music
withal further to ennoble it; but he must rise to
the level of his responsibilities.

The detailed discussion of this section has naturally
been confined to the English language, but it may
safely be said that the contention "Sing as you speak"
applies in the main to German or any other European
tongue.

(b) PROSODY AND METRE

If singing is to be speech in song, it must primarily
talk sense and talk that sense intelligibly. Here the
singer is for the first time face to face with opposition
from his friend music. So far, music has helped him.
It has demanded hard work from him, but when he
has worked it has backed him up. It has every wish
to continue to pay its share of the expenses, but it
cannot help itself; it has become, most unwillingly, a
passive resister. To talk sense intelligibly, the talker,
be he speaker or singer, must give to his words their
right values according to the rules of prosody. In our
language the variations of word-accentuation are very
numerous; beyond the natural metrical tendency to
throw the strong accents on the early part of the word,

there seems to be no settled rule to go by. To accommodate the prosody of his language to the rhythm of his music is the singer's difficulty, for music, with the best of good wills, is so hampered by the limitations of its own notation that it often acts as a positive drag on the wheel. In the old *bel canto* days, as explained earlier, this did not so much matter; sense and prosody did not count against beauty of sound and finished execution, and consequently were not so much catered for. But in the true art-songs from Purcell through Schubert to the present day, where the poem has been the original inspiration and the music its resultant outward expression, the text has been treated with equal honour, and the lilt of its language followed faithfully by the composer so far as the restrictions of his medium have permitted him.

These restrictions he will always have to face. For translating into song the countless inflexions of speech-quantities and accents he has a petty vocabulary of arbitrary semibreves, minims, crotchets and quavers with their subdivisions, and a mere handful of time-signatures with which to cope with metre. These he can supplement to a certain extent by stage directions, as a help or corrective as the case may be, but the fact remains that the text can never count upon untrammelled assistance from music in all the aspects of its expression. The text too must always remember that the music's the thing; the music must always come first, even in song, where speech would appear to be the most important. Rhythm is the beginning of all things, and its call is so powerful that it masters all else; therefore even the music, which is the outcome of the poem, takes precedence of the very text which inspired it. Wherever the reverse has been attempted, and the music plastically fitted with inelastic determination to the words, the result has been the same — the initial fascination of the experiment has palled in time for want of the great elemental force rhythm. The reconciliation of the two, text and music, is the difficulty. The modern musician has faced it squarely and risen to the occasion, as many of his predecessors did before him. Schubert seemed to have the power by nature. The singer can open any volume of his songs where he will and see it for himself. Let him take the first song in the first volume (Peters' edition), our old friend "Das Wandern," and read it through, text first, music afterwards. He will find that word by word, line by line, in prosody and metre alike, the text has not only been followed faithfully, but strengthened and ennobled in the doing. Let him move on to "Der Neugierige," No. 6 in the same series. Here he will find what is practically perfection of speech in written song. Thence, if he will, to the greatest song in the world "Der Doppelgänger." Could that most human and most moving of all tragic songs have stirred men's hearts like nothing else for the better part of a century if its message had not been truly *spoken?* Go where he will among them all he will find that same truth of expression, the pure translation of speech into song, which made that author the greatest song-writer of all time. The romanticists too felt it in their blood. Schumann's songs are models of the treatment of prosody values. The "Dichterliebe" and the "Frauenliebe und Leben" are such spontaneous speech in song that manipulation or reconciliation is practically a negligible quantity to the trained singer.

Their sentences run from words to music and from music to words with such emotional truth that we cannot think of them apart. Here is no waiting for florid passages to spin themselves out, no dragged-in vocal effects, no false quantities to fit the note at the expense of the word. We feel that he read the words and, as he read them, sang them for us. The ultra-romanticists of our own day have felt this compelling sense of words, and to its emotional drive we owe such purely rhapsodical songs as Maude Valérie White's "Absent yet present," which as a model of accentuation could not be bettered; or Hatton's "To Anthea," which in spite of its delightful virility and breeziness owes three-quarters of its charm to its faithful reproduction of word-values.

The modern composer, knowing the singer's difficulties, has done his best for him, and, be the metre what it may, has written his trochees, iambics, spondees and dactyls so that he who runs may read; that is to say, he has absorbed the metre of his poem and expressed it trochaically, iambically, or whatever it may be, to the best of his ability. But musical notation is limited. He has no means of insisting on, or even indicating, the longs and shorts of his prosody other than the natural accents of his rhythm, which he wants, as a matter of fact, for the modelling of his sentences, not for the scanning of his words. He can, it is true, convey the sense of his trochees by writing his song in $\frac{6}{8}$ or $\frac{9}{8}$ or $\frac{12}{8}$ time; but supposing he feels his song in $\frac{2}{4}$ or $\frac{3}{4}$ or $\frac{4}{4}$, what then? If he writes in $\frac{4}{4}$ time, he can dot his crotchets and give it that way. But supposing he does not want his crotchets dotted? Here we have arrived apparently at an *impasse*. His trochees (or iambics) are arbitrarily to be made into spondees, because musical notation only gives him two even crotchets to write them with. This is rather hard on the composer, not to speak of the poet. Surely it is about time for the singer to lend a hand. Why should those two crotchets, *written* of even value, be *sung* with even values? He does not *speak* them with even values. He may give them (though even that is not always necessary) equal *time*-values, but surely there is no necessity for equal *pressure*-values. He would not speak the words "pebble," "comma," "faster," as "pēbb-ūrl," "cōm-māh," "fāst-ūr," any more than he would say "rē-nēw," "āy-gāin," "prēe-tēnd," unless he were a prig. Yet in any song he may meet any of those words written to two equal crotchets or quavers. He has simply got to use his common sense and sing them with the same pressure-values as he speaks them. And this is not all; the minor words of his spoken language, such as the definite and indefinite articles, the prepositions and the rest, are no more provided with higher rank by being mounted upon notes in singing, than the trooper is made a colonel by being provided with a horse. Whatever values they have in speech they should be given in song — always on the understanding that, if ever there is a dispute in the matter between text and music, the music should have the preference.

In order to talk sense in song and to fulfil his share of the bargain, the singer has to ask a certain license of the composer. He must not be trammelled by hide-bound restrictions of note-values. Notes were made for man, not man for notes. If the singer is an interpreter, the composer will give him all he asks. What are note-values to him against the living song? He wrote them thus because he had no other way to write them. In exchange for one glimmer of the spirit he will give away the letter with both hands. Let us now apply that spirit to the letter of the Schubert song often quoted already, "Der Leiermann"; using for the purpose, as likely to make the most general appeal, the admirable English translation of Mr. Paul England.

Here is the poem as it stands:

"THE HURDY-GURDY MAN"

Yonder stands a poor old hurdy-gurdy man,
With his frozen fingers playing all he can,
Barefoot, shuffling sidelong on the icy way,
Not a single penny in his empty tray.

No one seems to heed him, no one stops to hear,
Only snarling mongrels care to venture near;
Little does he trouble, come whatever may,
Still his hurdy-gurdy drones and drones away.

Wonderful old fellow! Shall I with you go?
Will you drone your music to my songs of woe?

Let us now write it as we speak it, indicating the comparative values as follows:

(a) Pressure-values of syllables in the word, by longs and shorts.

(b) Importance of words in the sentence, by size of lettering.

"Yonder" implies a pointed forefinger (pointed probably mentally rather than actually). What does it point at? A man. A man in himself is no uncommon sight; this is a man with a hurdy-gurdy;

the hurdy-gurdy gives him his distinctive importance. Whether he was standing or walking does not matter; all that matters is that he looks poor and old.

YONDER stands a POOR OLD HURDY-GURDY man,

He is evidently bitterly cold. His fingers are so frozen he can hardly turn the handle. He does his best — all he can.

With his FROZEN FINGERS playing ALL he CAN,

Your eyes fall from his frozen hands to his frozen feet; you see — with a gulp in your throat — that they are bare. He keeps moving — probably in unconscious self-defence. You shiver as you think how icy cold he must be.

BAREFOOT, shuffling sidelong on the ICY way,

Not a coin of any sort — not even a penny does a single soul give him. (The assumed denomination and the singularity are equally pitiful.) He grinds and grinds away, and the tray is always empty. Presumably it is a tray — it may be a box or a cup, it doesn't matter.

NOT a SINGLE PENNY in his EMPTY tray.

He is utterly lonely. It is no wonder the tray is empty. Not a soul looks at him, far less stops to listen.

NO ONE seems to HEED him, NO ONE STOPS to hear,

Your eye is caught by a stray dog or two walking round him suspiciously. You can see in imagination their lips curling back and their teeth showing. It doesn't matter about their venturing near — they would not have caught your eye if they had not; the point is that they were the only living things that seemed to be aware of his existence.

ONLY SNARLING MONGRELS care to venture near;

Now comes the master-phrase of the song (*vide supra*, p. 17). You sing it as though you shrugged your shoulders for him.

LITTLE does he TROUBLE,. come WHATEVER may,

The drone obsesses your very senses; it goes on and on. Your eye cannot move from the hurdy-gurdy.

STILL his HURDY-GURDY DRONES and DRONES away.

How does he stand it? Who is he? What is his history? Does he feel anything at all? Why doesn't he die of cold or starvation or monotony of existence? He is truly a mystery. "Fellow" draws you a little nearer to him in spirit; "Shall" asks the question; "go" introduces a new idea — preposterous if it were not that you too are a hurdy-gurdy man, a poet. You are — or think you are — just as cold, just as dreary, just as miserable, just as hopeless a failure, and you ask him, under your breath, to take you into partnership.

WONDERFUL old FELLOW! SHALL I with YOU GO?

Will YOU DRONE YOUR MUSIC to MY SONGS of WOE?

Now let us fit these values to the music, or rather accommodate the music to these values.

(N.B. — The reader should, if possible, follow this with the actual song.)

The song is written in $\frac{3}{4}$ time. Two bars go to each line of the first eight lines. Each bar consists of either six quavers or four quavers and a crotchet. There are only four dotted quavers in the whole of the sixteen bars. The metre is trochaic. If Schubert had wanted to insist on his metre he could have written the song in $\frac{9}{8}$ time. This would have emphasised his trochees, but at the expense of his atmosphere — the old man would not have shuffled; he would have danced! Monotony is the soul of that atmosphere, and dreariness its very mood. The singer has no great rhythmic call to obey; the instrumental part takes that on its own shoulders — it is the embodiment of rhythmic monotony. All he has to do is to visualise his scene in his imagination, concentrate his hearers' attention on the *main* figures in the picture, letting all the rest go, *sound* monotonous, hopeless and dreary, and talk sense the while. If he sang those opening words with their full quaver pressure-values or even time-values, he would be monotonous with a vengeance, so monotonous as to draw his hearers' attention away from the song to himself — the very negation of interpretation. Once again the wood could not be seen for the trees — over-monotonising instead of over-elaborating, that is all. Schubert wrote those six quavers to the bar, because six quavers was the only way he had to write it without arbitrarily stereotyping a reading of simple words which might not in the least commend itself to another reader with other tendencies of direct expression. To pretend that he would have insisted upon their faithful reproduction is a poor compliment to the greatest master of song the world has yet known. Let us now suppose that, knowing the time to be $\frac{3}{4}$, we have a free hand to value its component notes our own way in accordance with our individual interpretation of the lilt of the words, only taking especial care to dot those notes which Schubert has dotted. This the skill of the translator has enabled us faithfully to do. It works out roughly as follows:

Etwas langsam. (*Rather slow.*)

Yon-der stands a poor old hur - dy-gur-dy man,

With his fro - zen fin-gers play-ing all he can,

Barefoot, shuffling side-long on the i - cy way,

Not a sin-gle pen-ny in his emp-ty tray,

Not a sin-gle pen - ny in his emp - ty tray.

No one seems to heed him, no one stops to hear,

On - ly snar-ling mongrels care to venture near;

K

Lit-tle does he trou-ble, come whatev-er may,

Still his hur - dy-gur-dy drones and drones a-way,

Still his hur-dy-gur - dy drones and drones a-way.

Won-der-ful old fel-low! Shall I with you go?

Will you drone your music to my songs of woe?

To reproduce the exact time-values in writing which the interpreter actually gives in singing is well-nigh impossible; the above is as near as can be got to this individual reading. The fine shadings of all such values are so intimate and variable, so at the mercy of mood, that they might never work out twice alike; no reading of any art-song can be arbitrary in any values of any sort. One thing can be stated with confidence. The listener may know his "Leiermann" by heart, he may actually follow in the printed page the performance of the above or any other closely allied version (or perversion) of the original, and he will never be conscious that a note-value has been altered! Not only this, but the modern composer will accompany his own songs and see them thus "man-handled" by the interpreter on the evidence of the

very notes before him, and probably never take it in.
In the jam of the spirit he will have swallowed the
powder of the letter unconsciously.

The modern composer has found out all this for
himself. The great men of to-day have risen so far
above the limitations of the written note that the
singer finds little left for him to do but sing the music
as it stands. He may unconsciously dot a note here
or there, or manipulate the pressure-values of a word
to suit his individual sense of sound; but there will
be no need for original enterprise on his part in this
particular direction. In fact it would sound out of
place and might possibly throw the song out of gear.
Hubert Parry's "A Lover's Garland" is a case in
point. Here is a positive gem of modern *bel canto*.
It has not got a dramatic point in it, its atmosphere is
carried on the face of it, and there are no contrasts to
make effects with; but it is the most perfect translation into musical language of the phrasing, inflexions
and word-values of the poem and its metre. Here
your longs and shorts are all made for you in the
swing of the music; you have nothing to do but sing
them. If you stop to try and equalise your crotchet-
values, before you know where you are the lilt of the
song has picked you up and hustled you along like a
bottle in a mill-race.

is a delightful example of the composer's cunning.
"Blossom" and "petal" are both trochees; but there

is a subtle difference in the time-duration of the first foot of each. The first syllable of "petal" has in speech a tendency to drop from its perch a little quicker than that of "blossom." That tendency the composer has put for you in black and white. He has taken the very bit out of your mouth, for you wanted to do that yourself! But because "petal" is a little different from "blossom" it doesn't follow that "blossom" is different from "petal" — comparatively speaking. Breathes there a prig with soul so dead who would speak of "blōs-sōm"? (If the composer had considered him normal, or even sporadic, he might no doubt have provided for him in anticipation with equal ease.) Through the whole song (as in the poem by Alfred Perceval Graves), with all its delightful modern spontaneity, runs a feeling of the classics, a beauty of outline, of Greek art, of the very spirit of the Greek anthology to which the poem owes its being. The word *bel canto* is applied simply to the style of its technique; the song is an art-song in its truest sense.

The modern *florid* song is practically non-existent. The florid songs of any period are too taken up with their own "floridity" to bother about prosody or metre; the two things are essentially incompatible. In the modern *rhythmical* song the singer will find it all ready-made for him, and, if he is wise, will let well alone. The composer is a far better judge of rhythm than he, and the rhythm is what counts. The singer may dot a note or put in a syncopation here or there, if thereby he can make his individual rendering of the rhythm more effective, but even such slight changes are generally made for his own convenience, not the song's benefit.

It would be possible to quote examples of this
divisional rule (b) ad infinitum. It applies to about half
of all songs and three-quarters of our language. The
singer must follow his common sense and fit language
and music one to the other where he can. To do it
he must be absolute master of his technique, and he
must feel the power of word-sounds. He must so feel
the music of their values that the spondaic treatment
of such words as "glory," "stature," "champion,"
"wrestle," "stately," will positively hurt; while the
giving of equal pressure-pomposity to each syllable of
such iambics as "veneer," "perhaps," "befall," or such
dactyls as "masterly," "canticle," "syllable," will move
him to laughter. There is no space to go further into
this question here.

In the free handling of his language the singer has
one great advantage — he is not a chorus. The idea
of manipulating a Plain-song or worrying the congre-
gational singing of "O God, our help in ages past"
with prosody values is unthinkable. The attempt to
dictate even expression values led to the perpetration of
hymn-books harried with ff's and pp's and to rank artificiality of sentiment in the singing of the great tunes;
but the hard-bitten Englishman is an anomaly. He
loves his treacly hymn-tunes, and he loves his treacle
thick. It is a question which he enjoys most, the
buffets of his Saturday football or the chromatic senti-
mentalities, Vox humana stops and dragged-out cadences
of his Sunday singing. The very essence of congrega-
tional singing, and of the chorale, is spontaneity, and the
warmed-up Wesley or bedevilled Bach is a sin against
purity. It is, however, a question whether a judicious
regard for prosody values could not often produce great

effects even in chorus work.   Take, for instance, the following passage from the famous chorus in Parry's *Job*:

The very meaning of the word *shouted* demands
trochaic treatment. If each voice in turn sings it as
"shōutēd," with equal syllabic values, not only does
the word lose its spontaneity and joy, but the force of
the penultimate bar, where such a spondaic broadening is not only legitimate but essential to the climax,
is anticipated and the climax itself weakened. The
elastic and responsive chorus probably does it for itself
by instinct. Whether it can be taught with safety
depends upon the chorus and the chorus-master.

There is no doubt that the English singer's — and in
some cases composer's — indifference to prosody values
is largely due to our cathedral traditions. The point-
ing of our Psalters is positively laughable; but custom
has so "staled their infinite variety" that choir and
congregation swallow the most ludicrous anachronisms
verse after verse, in chant and psalms, and never wink
an eyelash. Probably not one man in a hundred in
our churches when he sings Wesley's famous hymn
"The Church's one foundation" ("very slow" in *tempo*
as, we know from Sir Hubert Parry, Wesley wished it
to be), realises that in the word "Church's" he sings
a whole long beat on an apostrophe! or, worse still,
later on interpolates a vowel between the *schis* and the
ms of "schisms" — "by schīs-ūms rent a-sunder" —
giving it a whole beat to itself and exalting thereby
an appalling solecism into an accepted interpretation.
Long usage has made him callous, and his ear no
longer demands the verities of language; and the
singer, being asked for nothing, gives it.

But the singer has other active enemies in both
music and language themselves. The first of these is
the rising phrase. The rising phrase implies a natural

crescendo (the *diminuendo* and the dead-level rising phrases are generally written merely for purposes of word-illustration), and that natural *crescendo* gives a natural importance to its point of climax. If that climax happens to fall upon an insignificant word or short foot so much the worse for the word! The music's the thing, and the *crescendo* has the first call; so prosody must go. This is so instinctively recognized that the most fanatical prosody-purist feels no resentment. It is to be found all through song of any period. A couple of examples will do.

mein Aug' und   Herz,   mein Aug' und   Herz.

from Schubert's "Du bist die Ruh" is a case in point. Here the rising phrase gives a false value — increased by the dot — to the entirely unimportant word "und." It is true that the real climax is on the first beat of the next bar, but none the less the physical effect of the upward *crescendo* gives an anticipatory climax to the weak word even on the weak beat. The same is to be found on the *ner* of "deiner," and "es," in the other two verses. To exchange the pure *cantilena* effect of the passage for the sake of the word-values would be a poor bargain. The *crescendo* in such a passage is not marked, firstly because it is implied, and secondly because Schubert, like all great composers, was sparing with his expression marks.

The phrase quoted earlier (p. 72) from Brahms's "Feldeinsamkeit" also illustrates it. Here the words "sēndĕ," "lāngĕ," "ōbĕn," can be, and should be, sung with their speech-values; but the *nen* of "mēinĕn,"

though quite short in itself, is by sheer force of the
rising phrase promoted to an even more important
position than its long-footed brother, that short step
up of two notes in the social scale causing it to turn
its back on its friends of trochaic days. When the
singer is met by such glaring anachronisms of word-
treatment as the same composer's "Die Mainacht," in
which, as shown earlier (p. 73), most of the principal
accents come upon unimportant words, and a whole
throbbing bar with an inherent *crescendo* is given up
to the word "und,"

und   die ein - sa-me Thrä   .   ne rinnt.

he walks round it at first, like a dog round a cat, at
a respectful distance, before finally "sailing in." He
sails in at last because it is worth it. The beauty of
that tune and the depth of its emotional power swallow
up the defects. The composer must have known that
they were inconsistent, but he let them go rather than
disturb the lovely tune, and we must ever be grateful
that he did.

It might in passing be pointed out that in the matter
of syllabic values the German language is very much
like our own. The trochaic "kommen," "schönen,"
"saget," etc., rank practically like our "cometh,"
"lovely," "sayeth" — "saget" contracted often to "sagt"
like our "sayeth" to "saith" or "says." It is no uncommon thing to hear the first line of the "Dichterliebe" sung *Anglice* as follows:

"Eēm wōōndērschönāym Mōnāht Mā-ēē."

Here, to begin with, the word "Im" has not the *ee* sound at all but is far closer to the *i* in our word "dim"; the *u* of "wunder" is not "oo" but a rather broad version of the *u* in our "pull"; while syllabically the *der* and *nen* of "wunderschönen" are equivalent in value to the *der* and *ven* of our "wonder" and "given" respectively, and the *at* of "Monat" to the *ot* of "pilot," allowing for the brighter sound of the open vowel. The word "Im" is as unimportant as our "in." Here we have four words into which are generally crowded some five "howlers."

The same applies to the definite and indefinite articles. The English singer, in spite of his natural self-consciousness, tackles his German with a delightful absence of shyness. The thought that a colloquial knowledge of the language might help does not enter his brain. He sings his German song for his German audience with the same gusto that he sings his Irish song with a Brixton brogue in Dublin. ("Annie Laurie" is as inevitable an encore song for the Hampstead soprano in Glasgow as "Comin' thro the rye" in the Albert Hall in London.)

The next enemy is the *incidental high note*. It need not necessarily be really high; any skip that causes a certain physical effort entails automatically a certain accentuation. The accent on *with* in "If *with* all your hearts" from *Elijah* is natural enough — the skip from the B♭ to the high G necessitates it; but "Angels ever bright *and* fair," or "But *the* Lord is mindful of his own," are merely concessions to weakness; there is only a rise of a single note in each case, yet the unimportant word and weak beat are invariably inflated in value. "O ruddier *than* the cherry" is more excusable, for here there is a considerable skip to one of the higher notes of the bass voice; still it would be interesting to hear the song sung throughout with the face-values of the words and beats. In these cases the actual musical call is not so strong as in the rising phrase, but the physical demand for emphasis is even greater. It is better to give in to the demand, even at the expense of false emphasis, rather than cramp style; but there is no question that, as often as not, the note can be brought into line and given its right word-value with no loss to the power of the phrase. In all these cases, as in the whole application of this rule, the singer must use his common sense.

Next comes the question of translations. The trained English singer may have a fair knowledge of, say, German, French and Italian, and will sing the songs of those languages by preference in their original versions. But there are many languages, such as Russian or Hungarian, each possessing a wonderful song literature of its own in a language which he has had no opportunity of studying. Here there is no option for him but to sing translations. The essence of good song-translation is the giving of the poetic equivalent. Literal translation can never be anything but cramped in style and clumsy in technique. The true *singing* translation is freedom itself; it should be sound English, should be able to rank as a poem, and should *follow the lilt of the song.* It should stand on its own feet and not be hampered by limitations of accurate significance of words, provided only that it gives the atmospheric idea of the original. The freer the hand, the better the chance for prosody and metre. Only a singer can fully appreciate this; and on the rare occasions when

the translator is both singer and poet himself —
"Schuhmacher und Poet dazu" — we get masterpieces
to sing. In the case of a language like Hungarian
there is no way out of the difficulty. Here we have
the most marked rhythm that exists in music. We
know from Mr. Francis Korbay that that rhythm in
Hungarian music is the direct translation into sound
of the same rhythm in the Hungarian language, the
language being the *causa causans* of the music. The
two are so interdependent as to be practically one.
The Hungarian language and the English are as far
apart rhythmically as the poles; no manipulation of
our language, however free, could give an adequate
equivalent for the other and at the same time preserve
our own poetic prosody values. We have two alternatives. We can stop the grand march of that metre
and give the translator a chance — that would indeed
be burning the house down to roast the pig, or we
can frankly accept the situation, and agree that that
incomparable rhythm is worth to song all the language
values in the world. "Far and high the cranes give
cry," "Mohàc's Field" and "Shepherd, see thy horse's
foaming mane" decide it for us without even giving
us time to think.

Last, but not least, comes the singer's old friend and
enemy — the straight line in phrasing. How is he to
keep his poise, how is his song to run on telegraph
wires without a stop, if, by altering his syllable-values
for prosody purposes, he turns the continuous straight
line into a series of dots and dashes? But dots and
dashes are as much part of the message of the song as
of the message of the wire. Both one and the other
depend upon them for sense, and, strange as it may

sound, even singing can talk sense. In both cases that sense depends upon the skill of the operator. The singer who is not master of his technique can never reconcile the two — the straight line and the common sense.

He may be born with style and anxious for the fight, but, if he has not learned his business, the word and its meaning will fall before the magic of the phrase. That "stylist" is well known.

He may not know it counted; he may say to himself that the music's the thing, and the sense doesn't ever matter. That man is to be pitied; the public has never told him, and he will never learn.

He may say to himself that the voice is the thing, and the sense may go to the wall. His career will be short and cheap. *Ars longa, vita brevis.*

He may, *more Anglo-Italiano,* carry over and conjoin, and, having fulfilled his whole duty to both, bury his head in the sand. It is not so ostrich-like a manœuvre as it appears; it is the selling of a patch-work quilt for a Persian rug.

Or he may have taken off his coat at the beginning and learned his business like a man. Learning your business is not a drag on the wheel of genius. Hans Sachs has told us that. If song makes the greatest emotional appeal from man to man, surely it is worth the singer's while to learn its language. To the master of style who is also the skilled workman, the dove-tailing of the two — phrasing and sense — is a labour of love.

(c) IDENTITY OF TEXTURE

Every song has an atmosphere of its own; it also has a *texture* running through the material of its composition. Such a texture in the written song is inherent and literary; its preservation in the outward and vocal interpretation is part of the singer's duty. To deliver it truly outwardly he must, naturally, first appreciate it inwardly. If he is a master of style, texture will be inseparable from the song as a whole, and patchiness equivalent to cheap workmanship; but such inward knowledge must be supplemented by outward application in order to make the song, when sung, a consistent whole. Identity of texture between the singing and the speaking voice is essential to such a true expression of the literary text, and incompatible with patchiness in its musical interpretation.

The singer who experiences difficulty in singing a certain song or phrase with pure diction, will find it an admirable plan to choose out some one note in his voice which gives him no effort to sing, and with whose quality he is satisfied, and on that note to *intone* the line or poem in question. The difficulties which bothered him before were probably vocal, not concerned with diction, and the transference of the subject to another medium causes them to disappear temporarily. Worries of voice-production and uncontrolled breath-pressure have thrown him off the line in the song, and with them purity of diction has suffered meantime. On the intoned note these fears are absent, and he has no long phrasing to disturb his thoughts; he consequently has time to think about his words, and pay attention to the common sense of their sound and their texture after transplanting them into new surroundings.

When, therefore, he intones at ease, he must see to it that every word he *sings* shall have *exactly the same quality of sound or texture* as in *pure speech,* speech as near the speech of ordinary conversation as is consistent with the rules of pure oratory. If he detects the slightest difference, let him first speak the line word by word, and then sing it, and repeat the process again and again until it is right. Then, when his ear is not only reconciled, but devoted to their agreement, let him transplant the words, texture and all, to the song and repeat it there till the new reading is his own, and he has not got to stop and think about it. The sense of power he thereby acquires will surprise him. Audiences which lazily applauded him from good-nature or for the beauty of his voice, will suddenly begin to attend to him, for he begins to talk to them in a language they understand. Speech in song is the most moving of human gifts. The singer who goes out to sing, and leaves his speech behind, is like the man who went out to shoot with powder, shot, wads and caps, and forgot the gun.

The English singer is the most healthy minded and probably the most gifted vocally of any. If he would but appreciate his responsibilities and opportunities, if he would but widen his horizon, he might at any moment step out of provincialism and take his place in the world. The ability to paint a Christmas card or chocolate box with tolerable success and quick returns is no guarantee of the achievement of a masterpiece of portraiture. A beautiful voice is a beautiful voice and no more; the clapping of hands is but the clapping of hands. The one may meet the other of an evening and coo the language of suburban love; but the offspring of that union does nothing for its country and is generally degenerate.

Those who have followed the rules set forth in this Part II. will have noticed how interdependent those rules are, both in their working and their appeal. The appeal is temperamental to the hearer's primitive instincts and balanced reason; the means are *rhythm* and *language*. If rhythm is horizontal in direction and has for its motto "Push on," then a song must never *stop*. If language is to be as intellectual in its appeal as rhythm is physical, that language must be the language of both singer and hearer, not a compromise of vulgar anachronisms and cowardly avoidances.

If the song is to be treated intellectually as a whole, and if that song is ever on the move, then the intellectual singer must be ever on the move with it. Through every bar, sung or unsung, he and his language must sing, live, move and have their being.

Let him finally remember that his physical application of these rules — the *grip* of Rule I.; the *ready, aye ready* of Rule II.; and the *fluency* of Rule III. — depends upon his appreciation of the fact that what counts is not the large amount of breath which he takes in, but the small amount which he gives out.

PART III

MISCELLANEOUS POINTS

STYLES OF TECHNIQUE

THERE are, roughly speaking, three styles of technique; that is, three styles of physical application of the voice to the expression of the meaning of the song.

(1) *Bel canto*, in which beauty of sound and pure singing are the first consideration, and words but the medium for conveying them in their most intelligible and sympathetic form to the human being. Under this head come about three-quarters of all songs and singing. This is as it should be, for the fundamental idea of song is beauty of sound and the ennobling of language. *Bel canto* technique has none of the limitations of the *bel canto* school of song referred to earlier, but embraces the whole world of song, from the most stereotyped florid aria to the most modern work of art. Whether it be Giordani's "Caro mio ben," Bach's "Todessehnsucht," Handel's "Rejoice greatly," Schubert's "Abschied" or Hugo Wolf's "Anakreon's Grab," pure singing and beauty of tone and phrasing are vital to the interpretation of one and all of them. Every song may be assumed to require *bel canto* technique unless it specifically demands one of the other styles.

(2) *Declamation*, in which dramatic expression is of
paramount importance, and the voice and its colours
are used as the most dramatic means for that expression; where actual beauty of sound is subordinated
to serve its purpose (mostly as a contrast) in the
general scheme of the particular interpretation. One
of the commonest illustrations of this is the *declamatory* recitative followed by the *bel canto* aria, where
the incisive materialism, as it were, of the one serves
to enhance the smooth idealism of the other when it
comes. The great dramatic songs are mostly declamatory either in whole or in part, such as "Der Doppelgänger," or "Der Erlkönig," of Schubert, in which,
though there are bars of *bel canto*, yet the interpretation
depends upon dramatic declamation; or Schumann's
"Waldesgespräch"; or Korbay's "Mohàc's Field"; or
such a frankly murderously-minded masterpiece as his
"Shepherd, see thy Horse's foaming Mane." In such
songs strength and virility are the all-compelling forces,
and their subordination to strict beauty of sound would
not only convey an impression of effeminacy, but would
actually damage the character of the song in the interpretation. Sung by a beautiful voice declamatory singing should always be beautiful; but strength, rhythm,
incisiveness and dramatic illustration should be the
singer's chief concern.

(3) *Diction*, in which the *words* come first, and the
music is but the medium of expressing them most
effectively. To this style belong most of the songs
en miniature referred to earlier (p. 77). They are
generally quick in *tempo*, and do not convey any very
deep emotion, except in its rather ecstatic form, such
as "Die Rose, die Lilie" from Schumann's "Dichter-

liebe." They are, as a rule, happy or humorous or lively in sentiment. Schubert's "Haidenröslein," Schumann's "Ich kann's nicht fassen nicht glauben" (from the "Frauenliebe und Leben"), Brahms's "Dort in den Weiden" and "Vergebliches Ständchen," Ernest Walker's "Corinna's going a-maying," Stanford's "Did you ever?" "The Crow" and "Daddy-Long-Legs" (from the song-cycle "Cushendall"), Hubert Parry's "Follow a Shadow," A. M. Goodhart's "Mary" and "The Bells of Clermont Town," are typical examples; while folk-song is full of them; such as "Der Kukkuk," "Spinnerliedchen," "Ecoute d'Jeannette," "Trottin' to the Fair," "Quick! we have but a Second," and hosts of others. In all these beauty of tone need not be absent, but tone as a first consideration would clog the wheels both of the movement and the sense, and defeat interpretation.

Many songs contain more than one of these styles — in the "Erlkönig," for instance, or "Waldesgespräch," mentioned above, we find, in juxtaposition to the declamatory, the pure *bel canto* of the "seducing" voice of the Erlking and the Lorelei respectively; in fact, the dramatic effect of each song depends upon that very intimate contrast. In "Der Neugierige" of Schubert we have three bars of diction ("Ja, heisst das eine Wörtchen, das andre heisset Nein") in quasi-recitative, interpolated into the middle of a pure *bel canto* song with delightful effect; while in such a song as Korbay's "Shepherd, see thy Horse's foaming Mane," the opening questions are as unmistakably *diction* as the answers are *declamation*.

There is one other style of technique — perhaps the most interesting of all — that which combines *bel canto* and *diction* in one and the same sung phrase. This is used in such atmospheric songs as "Der Leiermann," where the tone must be as beautiful as is consistent with inconspicuousness; and in those introspective deeply emotional moments where, in the anaesthesia of the tragedy, all consciousness of the outside world seems to be lost, such as "Nun hast du mir den ersten Schmerz gethan" from the "Frauenliebe und Leben," or the opening bars of "Der Doppelgänger." The stunned impersonality of the one and the hypnotic trance of the other, when truly *spoken* by the emotional singer, are among the most affecting things in music.

The aesthetic diagnosis of the song and the choice of its technical treatment are the privileges of the interpreter.

THE SINGING OF RECITATIVE

Recitative starts handicapped by three popular misapprehensions.

(1) That it is to be sung either entirely *ad libitum* or strictly foursquare; the choice depending upon the disposition, assertive or retiring, as the case may be, of the singer.

(2) That its style of technique is always declamatory.

(3) That nothing is recitative that is not either specifically marked as such, or so plainly indicated as to be unmistakable.

(1) In recitative the singer has the stage to himself. That is the privilege which it confers upon him, and with it often goes stage-fright. With no accompaniment to give him a lead, he either gets completely out of hand, or never lets go of his strict note-values.

The one is as disturbing as the other is dull.   Recitative is no more *ad libitum rhythmically* than a florid song; where the singer has freedom is in note-values. Recitative is written by the composer in those note-values in which he himself would sing it.   In most cases the composer's version is the best; but many singers would not feel the declamation in exactly the same way, and they have a perfect right reasonably to adapt the written note-values to suit their own histrionic ideas.   But that privilege carries its responsibilities.   They may do what they like in the bar, but the *bar itself must preserve the rhythm of the song.*   The physical effect of recitative is to stop movement, but recitative, like the song, must never stop in its march, except for definite dramatic purposes.   In *recitativo secco* this is hard to keep in mind, for the accompaniment seems always to wait upon the voice, and how fatal to the singer's balance that is has been shown already.   If the singer of the Evangelist in Bach's *St. Matthew Passion* forgot for an instant the inherent rhythm of his bars, the part, instead of being one of the most moving things in song, would sound interminable.   Let us apply the license to the following short recitative from Schubert, "An die Leyer," and try to realise the responsibility :

RECITATIVE.

Ich   will   von A - treus' Söh - nen,   von

Kad - mus will ich sing - en.

Schubert's values are really sufficient unto themselves
and cannot be improved upon, but there is no reason
why a singer should not sing the passage as follows:

Ich   will von A - treus' Söh-nen.

if that be his individual way of hurling his challenge
at Eros. But though he may handle Schubert's note-
values as he will, Schubert's time-signature must be a
law to him; the rhythm of the ¾ bar must ring in his
ears throughout. Looking at it superficially, he may
say, "What about my contrasts? If you shackle my
challenge with rhythm, you anticipate the submission
implied in the rhythmic *bel canto* which follows."
Schubert has seen to that for him. Every bar illustrates the struggle of the harpist to stick to the heroic
vein. Bar after bar it dominates the great rhythmic
chords of his lyre. Bar by bar it fails him, softens
down, halts in its stride and finally pauses — he can
sing of Love alone; the submission is complete. Then
follows the beautiful rhythmical air, one of the most
beautiful that even Schubert ever wrote. Had the
recitative failed in its rhythm, it would have anticipated
the very failure of the heroic mood which Schubert
has portrayed in the accompaniment, while the *diminuendo* and its histrionic illustration would have miserably
"petered out" together. The melody then, when it
came, would have started handicapped with anaemia
and itself have perished ignominiously.

In *accompanied* recitative the singer has his rhythm
provided for him. The wise conductor, though giving
him every license in spreading out his phrases, will

see to it that the phrases themselves "push on."
Many of the Bach accompanied recitatives, such as the
bass "'Twas in the cool of eventide" and the alto
"Ah! Golgotha!" from the *St. Matthew Passion,* are
practically Ariosos and can be treated as such; in
fact, if the singer will look upon all accompanied
recitative as a form of aria singing he cannot go far
wrong. Such recitative is deeply emotional as a rule,
and often more direct in its appeal than the florid air
which follows. It carries with it the old liability to
"foursquareness" in note-values and the consequent
lifelessness inherent in the slow *tempo.* If he will (in
pursuance of his Rule III., section *b*) give, whenever
possible, the prosody values of his speech to the note-
values of his music, the slowest recitative will live and
move, and the appeal will be the nobler for the giving
of understanding to the people. He does Bach more
honour by translating him with the truth of the spirit
than with the hyper-conscientiousness of the letter.
Such phrases, for instance, as the following from "Ah!
Golgotha!":

when sung with these note-values:

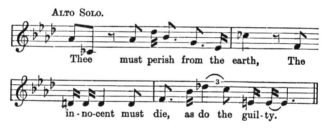

while the accompaniment remains the same, not only
do no hurt to Bach, or stop the flow of the rhythm,
but by their direct appeal to the intelligence through
language make the message clear. It cannot, of
course, be printed thus; no editor would thus arbitrarily alter Bach's original note-values. It is a
privilege of the singer, unwritten — and probably
unnoticed !

In this connection the following passage from Dr.
Henschel's book *Personal Recollections of Johannes
Brahms* does not come amiss. Brahms says, "As far
as I am concerned, a thinking sensible singer may,

without hesitation, change a note, which for some reason or other is for the time being out of his compass, into one which he can reach with comfort, provided always the declamation remains correct and the accentuation does not suffer." And again in a letter to the author, he says, "As far at least as my experience goes, everybody has, sooner or later, withdrawn his metronome marks. The so-called elastic *tempo* is moreover not a new invention. *Con discrezione* should be added to that as to many other things."

Here we have it on the very highest authority of our time — *con discrezione* in that and *many other things*. A legacy of trust to the "thinking and sensible singer." That singer does not ask to change a note; he merely asks to sing *con discrezione*, that his singing may be speech in song. Note-values are changed of necessity every day with and without *discrezione* to meet the exigencies of translation; what is license to the translator is not licentiousness to the interpreter. If the world will grant Brahms's *con discrezione* to the singer as a right, and the singer will accept it as a trust, enterprise and orthodoxy need never quarrel again.

It need hardly be said that there is no intention here to re-define recitative. What has been said is for the singer's ear alone. *Recitativo secco* must always be free as air — in opera the exigencies of dramatic action positively demand it — but if the composer for form's sake has written his bars so as to scan, the singer must sing them so as to move. Rhythm here, as everywhere in song, must be horizontal; and though the conductor may, quite rightly, accompany *recitativo secco* with down-beats alone, the notes that come between those beats should run along the wires. It is an appreciation of rhythm much too delicate to be dogmatically laid down as a rule. It belongs to the more intimate senses of interpretation and can only be suggested in black and white.

(2) These conductor's down-beats have a peculiar effect upon the singer. The relapse into the vertical seems hypnotically and automatically to change the style of his technique, and the sense of the text and run of the phrase go down before the convention of the word "recitative." In the passage immediately preceding the aria, "It is enough" from *Elijah*, Mendelssohn marks the recitative at the words "Tarry here, my servant." The recitative really begins, of course, with the words "Though stricken, they have not grieved," and the word "recitative" is only written where the *recitativo secco* begins and the singer is given a free hand. The feeling from beginning to end of the passage is one of pathos, and the style pure *bel canto;* yet it is a commonplace to hear the *bel canto* of the accompanied passage turn suddenly to the *declamation* at the first beat of the *recitativo secco*, making "hay" of the sense of both words and situation. One has only to compare this pure *bel canto* recitative with any of the "Baal" recitatives in the same work, to see what a wide range of technical style is possible. The latter are dramatic enough to be declamatory with a vengeance, but the one does not necessarily entail the other. The Evangelist in the *St. Matthew Passion* tells the greatest story of the world in incomparable dramatic form; yet every note is written and sung *bel canto* with deep and tender emotion. Even the earthquake recitative is far removed from "Take all

the prophets of Baal" in declamatory style. Practically all the Church cantata recitatives are *bel canto* in technique, and, be they accompanied, or *recitativo secco*, should be sung almost lyrically. The recitatives of Christ in the *St. Matthew Passion* are impossible to think of as declamatory. A large part of these is accompanied, it is true, but a large part is *recitativo secco* and in structure identical with the others; but the very idea of declamation is opposed to the whole personality of Christ and Bach's most loving picture of Him. The expression of that personality is truly human, but in its sorrows and its love for mankind rather than in its rebukes and its challenges. Even "Put up thy sword into its place" must contain the forgiveness born of knowledge.

Recitative singing, like all the other forms of interpretation, is governed primarily by common sense. The word "recitative" carries nothing with it but freedom, and principally freedom of choice. It arbitrarily demands nothing. The style of technique must be adapted to the common sense of the words and the style of their setting — the singer must cut his coat according to his cloth. In many cases recitative has been used by the composer simply to cover the ground. To read "drama" into such commonplace statements of fact as "And he journeyed with companions towards Damascus" (from Mendelssohn's *St. Paul*) and the like, is an excess of zeal permitted only to contraltos and tenors.

(3) Recitative directly described, or unmistakable, as such is practically confined to opera and oratorio; the printed word "recitative" in the Lied or song with pianoforte accompaniment is rare. Yet such songs

are full of it. It may not be indicated by any special
marks; none the less it is recitative in feeling and must
be treated as such. "Ich hab' im Traum geweinet"
from the "Dichterliebe" is recitative pure and simple;
so is Schubert's "Der greise Kopf"; also the beginning
and end of "Der Wanderer" obviously; "Der Leiermann" and "Der Doppelgänger" are practically recitative throughout, the one atmospheric and the other
dramatic, with their corresponding styles of technique.
Compare them with the pure *bel canto* "Das Wirthshaus," and the sense of recitative will materialise in
the singing. In "Ihr Bild" we have the two styles
in one and the same song, the actual effect depending
upon their juxtaposition and contrast, and the resolution,
as it were, of the first phrase in the second. Such
recitative is of course only implied, and its recognition
might be almost described as esoteric. It is for the
interpreter to feel it in his inner consciousness and
treat it accordingly. The detached sentence or phrase
implies it to a certain extent, but there are many pure
bel canto songs which carry the feeling of recitative in
them, and which, though bound by the rules of *song-
singing*, must convey the atmosphere of recitative to
the singer's mind. Such are "Wenn ich in deine
Augen seh'" from the "Dichterliebe," or "Nun hast
du mir den ersten Schmerz gethan" from the "Frauenliebe und Leben"; or, in the dramatic class, such a
mood-song as "I hate the Dreadful Hollow" from
Somervell's "Maud." Examples could be quoted *ad
infinitum*.

There are numbers of analogous instances in instrumental music. The introduction to the last movement
from Beethoven's String Quartet in A minor, Op. 132

(*più allegro*), is frank recitative. A more subtle example is Schumann's "Abendlied" (best known in the form of a violin solo), which has probably been looked upon as the purest piece of *cantilena* in music, and which nevertheless is pure recitative from beginning to end.

The thing for the singer to remember in recitative singing is that he has the stage to himself; the accompaniment is more or less but limelight. The independent and interdependent handling of dramatic recitative and instrumental illustration were shown us first by Gluck and apotheosised later by Wagner. In these the singer must take his appointed place. But in the main he has the privileges of a freedom which makes his responsibility all the greater.

PAUSES

THE Pause, or *fermata*, and its uses and abuses have been referred to earlier, but it is so effective when properly handled, and so dangerous when played with, that it deserves a chapter to itself. For convenience' sake it and its kindred of the *rubato* family — *tenuti* — will be referred to as "pauses" throughout.

The composer who knows his business is chary of his pauses. He knows that they stop the march of his song, and therefore writes them only where he knows they will enhance dramatic effect or strengthen the rhythm.

For the singer such pauses may be divided into two — pauses on notes, and pauses on silent beats. Pauses on notes are particularly dangerous to him because he unconsciously makes them upon effective notes in his voice, irrespective of their bearing upon interpretation. The trashy modern ballad is full of such pauses so distributed and manipulated, regardless of aesthetic meaning, as to give the singer the best chances of cheap vocal effect.

The written note-pause of the great composer must be treated with the same respect as his written note; the unwritten pause is the danger. Such unwritten pauses are in the nature of broadening of the rhythm, and in the proper place are invaluable. Their use and

158

abuse are emotional and individual to the singer. The spontaneous emotional pause is as effective as that made in imitation or to order is dull; one enhances rhythm, the other stops it. This applies not only to soloists, but even to a chorus. For example, a slight pause, or broadening, of the first note of the soprano lead

O may we soon a-gain re - new that song.

from Hubert Parry's "Blest pair of Sirens," when sung by a chorus possessing a soul of its own, will not only help the lilt of the phrase, but, as it picks up the rhythm, will convey the fascination of a deferred *tempo primo;* whereas a similar pause made by the "drilled" choir will sound an outrage upon the composer. Such an emotional use of the pause is too elastic and too intangible to put into words; the composer would certainly never write it.

Brahms's "Wie bist du meine Königin" contains an excellent example of the elastic pause. The poem (a translation from the Persian) is a poor one, and the song is given its emotional power by the beauty of the music; but the centre of the interest is in the treatment of the word *wonnevoll.* The composer has written it in each verse quite simply, without any expression marks; to pretend that it should be sung foursquare and unvaried each time is absurd. The singer's "pause" treatment of the passage in the individual verses cannot be even suggested; he probably could not suggest it himself. He may sing it differently every time he sings the song, unconsciously broadening or narrowing each phrase in response to the mood of the moment; if he is master of his technique, the voice will follow of itself. One thing is sure — he will not pay any attention to the strict duration values. He will sing it as he pleases, and not a soul will know he has taken any liberties at all. The very word "wonnevoll" is so ecstatic in its admiration — we have no equivalent for it in English — that it rises superior to crotchets and quavers, but none the less the song is pure *bel canto* in technique assimilated to purity and idealism in expression, and "points" and *ad captandum* effects are as foreign to its spirit as the shackles of notation to its letter. The sequence of "wonnevolls" may be alternately *a tempo*, long and short, or short and long, worked up by a *crescendo* or fined down by a *diminuendo*, but one and all must be in the spirit of the verse to which they belong, and one and all must be imbued with the rhythm of the song.

The "development" pause, or broadening of the repeated phrase as an enhancement of rhythm, has been illustrated earlier (p. 74) in Vaughan Williams's "Roadside Fire." Another example is the following passage from Elgar's "Sabbath Morn at Sea" (Sea-pictures, No. 3):

Here, again, the composer has marked no pauses. Nevertheless to sing the four phrases alike would be ridiculous. Such development of the natural *crescendo* of the phrase is so inherent in the repetition, that it would be unnecessary to reassert it, were there not many people who still feel themselves bound by the strict letter of the composer's directions. Such pauses — or rather *tenuti* — are not in reality pauses at all. They are in the nature of a spreading of the arms to rhythm, and are often accompanied on the stage by that very gesture; they gather up the threads of attention and hold an audience momentarily in suspense. They are an aristocratic version of the rush to the footlights, and the temporary cessation of nutcracking in the gallery. As in "The Flowers that bloom in the Spring" they "hold up" the song, turn out its pockets, and send it flying along the road of *tempo primo*. Compare the *marked* pause and *tempo primo* in Schubert's "Morgengruss" and (*en miniature*) "Haidenröslein," and the pause (unmarked but unmistakable) on the word *immerhin* in "Die junge Nonne."

Next comes the *dramatic* or *illustrative* note-pause, probably not written by the composer, but made by the singer either for purposes of dramatic illustration or to supplement the mood of the story. The "Sands of Dee" by Frederic Clay contains a very good instance. The song is a dramatic ballad and like "The Twa Sisters of Binnorie" is strong enough to bear illustration. The very repetition of the words "Call the cattle home" entitles the singer to regard their musical setting as the actual call to the cattle. A pause on each "call," as follows:

M

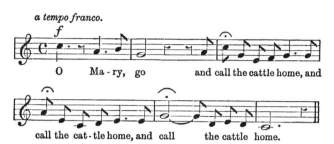

gives a verisimilitude to the cattle-call which the notes
without the pause would not so strongly convey.
Then, when in the last verse the same call is repeated
in "ghost" voice, the effect is increased a hundred-fold.
The composer did not write these pauses. Either he
did not think of them, or possibly he trusted to the
singer to make them for himself, or most probable of all,
he wrote them, remembered the British "popular" singer
and scratched them out again. Take, again, the pauses
on the main accents in Korbay's "Mohàc's Field."
Some are written, some are not; each one is illustrative of the mood of the song, of the defiance and virile
fighting spirit which dominate the whole. With these
may be classed such frankly illustrative sounds as
"echoes," where the singer must be given law in order
to achieve the effect.

The "question and answer" pause has been referred
to earlier (p. 82) in Stanford's "Broken Song." The
pause on the end of a question seems to be recognised
as music's natural expression of the question mark,
and the singer will find it in most cases written
for him wherever it is appropriate to the spirit and
tempo of the song, *vide* Schubert's "Der Wanderer"
(*immer Wo?*), which combines both question and

echo in one. If he does not find it so written he is
at perfect liberty to adopt it for himself, provided
that thereby he enhances effect and does not stop
the march. Wherever — as in some cases of question
and answer — the song is built of detached dramatic
sentences, he is entitled to handle it as he pleases.
Songs of "reminiscence" have a natural tendency to
pauses, inherent in the spirit of their atmosphere; the
voice, like the memory, loves to dwell upon the days
that are no more.

Lastly come the pauses on notes (1) immediately
preceding a return to the original subject, which are
but variations of the deferred *tempo primo* referred to
above; (2) little pauses or rather "pushes" upon
certain notes in a phrase, such as

Ruhig, zart (quiet, tender).

Ich sen - de ei - nen Gruss wie Duft der Ro - sen,

ich send' ihn an   ein Ro - sen - an - ge - sicht.

from Schumann's "Ich sende einen Gruss," which are
simply the little idiosyncrasies of style; and (3) pauses
for purposes of word-illustration, which will be referred
to later.

Pauses on rests.

Rests are not pauses; the song marches, and the
singer sings mentally, through them. Pauses on rests

are made for a definite purpose; to enhance one of the most telling effects in interpretation — dead silences. The composer knows their value well and their significance. Where the great composers have marked a rest-pause, its observance is practically compulsory. It has nothing to do with vocal effect, for the voice is silent; it is evidently vital to the interpretation. Its meaning is generally psychological, its very negation of action limiting its illustrative possibilities. Like all dead silences it tightens the strings of attention, and on the mind thus attuned the master-hand can play as he will. Its best known form is probably the pause at the end of the opening symphony before the voice begins. Such pauses, marked and unmarked, are a commonplace in the "potboiler," being "thrown in" to give time to the singer to pull himself together, and the audience to look at their programmes. The emasculated prelude which generally does service for the opening is equally effective with or without them. In a song like Schubert's "Die Stadt," however, one feels the pause's full significance. The opening symphony is dramatically illustrative — of the rowing of a boat — and pregnant with tragedy. The composer has marked a long pause at the end of it. During that pause the audience, stirred by the opening, holds its breath. The dead silence holds it spell-bound as nothing else could. Then out of the middle of that silence the story begins. Such silences are of the very essence of drama; they are well known on the stage, and, when they come, they hold attention in a vice. They are intimately associated with magnetism and correspondingly dangerous. Let them be held a second too long, and the audience given time to breathe — the thread snaps, the fixed eye wanders, and the song collapses like a house of cards.

Where the song is strophic and the pause is written before each verse, the intention of the composer is generally the same throughout; but where such a pause first occurs (marked or unmarked) in the middle of a song, it generally implies a change of sentiment and probably a corresponding change of colour. For instance, a pause before the last verse of Charles Wood's "Ethiopia saluting the Colours" is indispensable. The regiments have gone by, the old woman has disappeared, the day is over and the story is done. The last verse is a meditation pure and simple, and the "narrative" has given place to the "far away" colour of reminiscence. In that pause the speaker must have seen the last soldier disappear, watched the old woman out of sight, turned back to his tent, eaten his dinner, and lit his pipe.

The actual dramatic pause is well illustrated in the "mad" song — "Dead, long dead," from Arthur Somervell's "Maud." Here the pauses are inter-paroxysmal, and are used illustratively to separate the moods, the contrasts being thrown into relief by the silences between.

There is a pause known in German as the "Kunstpause," or "artistic pause," which cannot be definitely counted as either a "note" or "rest" pause, because it can come on either. It is a pause, generally of infinitesimal duration, before a given word, made for purposes of emphasis (generally quite gentle); a miniature stopping of the rhythm to point a finger. (See the reference, on p. 69 above, to the word "asleep" in Stanford's "Fairy Lough.") It has as antithesis the infinitesimal anticipation of the beat in strongly rhythmical songs, used to push the rhythm on more strenuously, if possible, than before. Both are delightfully effective, but must be used sparingly, for they degenerate quickly into mannerisms. The "dynamic" pause made for purely display purposes, though legitimate enough in its place, is of no interest, and need not be discussed here. The "cadence" pause for the ending of songs has also been treated of earlier (p. 81).

There is one other pause which, however, must not be practised by beginners. It might be called the "magnetic" pause, as its purpose is outside actual interpretation. It often happens that even the most magnetic singer may fail to hold the attention of his audience. If he has a pause in a legitimate place in the song — *both text and music must justify it* — let him convert that pause into a dead silence and hold it. As the silence makes itself conspicuous, all eyes will return in very curiosity to the platform, the indefinable fidgeting will melt away, the same dead silence will communicate itself to the room, and the singer with the reins of attention in his hand can start fair afresh.

The danger of all pauses is that they stop movement. The great composers knew it and know it. Schubert wrote no pauses in "Du bist die Ruh" or "Das Wirthshaus," or in any other song where they were not essential. In the whole *bel canto*, florid and rhythmical school they are rare, and chiefly used to enhance rhythm by the temporary "hold up." When he can use, and how long he can hold, his pauses is known to the interpreter by his sense of touch.

RUBATO

In musical language *rubato* is another word for
elasticity of phrasing, and, as such, appropriates most
of the credit for elastic interpretation. As a matter
of fact it has no merit of its own whatsoever; it is
simply the *agent provocateur* of rhythm. The charm
of the Chopin *rubato* is not due to any elaboration or
elongation of any group of notes, but to the return to
the *tempo primo* when the elaborating process is over.
It may, as in singing, be helped by change of colour,
but inherent virtues of its own it has none. The
concert-goer has unconsciously experienced this many
a time. Some famous pianist has played a famous
Nocturne with his famous *rubato* and has left him
cold. Why? Because the search after *rubato* carried
the player so far away from his fundamental rhythm
that when it was found again, both player and listener
had forgotten all about it. If the *rubato* had been
sufficient unto itself, he could not have had too much
of it: but *rubato* alone, as the listener knows to his
cost, is Dead Sea fruit. Its misunderstanding has
been responsible among other things for the degeneration of the Hungarian dance as played in England.
Cause and effect have been jumbled up, *rubato* has been
promoted and rhythm degraded, and the resultant
hotch-potch swallowed wholesale everywhere to-day for
"Hungarian as she is played." Perhaps the concert-
goer is still sceptical and requires further proof? Has
it ever occurred to him to ask why the more subtle
rhythms such as $\frac{5}{4}$ and $\frac{7}{4}$ are practically invariably
played without *rubato?* It is not from any difficulty
of execution; $\frac{5}{4}$ is as easy to play or sing as $\frac{4}{4}$. It

is simply because here *rubato* does not pay. Such rhythms to Anglo-Saxon ears (this does not apply to those nations which have quintuple rhythm in their blood) are not compelling. They owe their charm to their unconventionality not their strength; there is no insistent demand for return to the *tempo primo*, and *rubato*, therefore, is but a waste of time. Take, again, an old song like the sixteenth century "Ein schön' Tageweis" with its alternating $\frac{4}{4}$, $\frac{3}{2}$ and $\frac{6}{4}$. Such a fascinating lilt laughs at *rubato* and turns it out of doors. There is no room for it, because every other bar is either the *rubato* or the *tempo primo* of the one before.

Rubato ("robbed") by its very name implies friendly spoliation; its strength lies, however, not in the taking away but in the giving back. To the big predatory *rubato* family belong the *rallentando, accelerando, fermata, tenuto* and all the like individuals that have been spoken of earlier. No more bloodthirsty band of young cut-throats ever waylaid their mother in the passages and rifled her pockets at the point of the sword; yet as she tucks them up in bed at night, and stows away their lethal weapons for the morrow, she thanks Heaven for a happy day. What they robbed her of they gave her back a hundred-fold.

The singer may take it as an axiom that the success or failure of his *rubato* is in inverse ratio to his distance from his fundamental rhythm. The closer he holds to that, and the smaller the radius of his *rubato*, the more effective will be his interpretation. He is in far more danger than the instrumentalist, for the latter has music alone for his medium and, if he will have extraneous help, must appeal to platform eccentricities;

whereas the singer in addition to the music has the words with their potentialities for evil as well as good. Over-elaboration, over-dramatising, over-sentimentalising and cheap effect hide in the dark places, and each one holds a pistol at the head of rhythm. To join that band is to embark on a life of sin, and ultimate degeneration is but a matter of time. Rhythm will always be his best friend.

> "We fell out my wife and I,
> And kissed again with tears."

Let him stick by her for better or for worse, and when he leaves her let the parting be short.

CARRYING–OVER

Carrying-over, or the *joining-on* without a break of one subject to another, is a matter for the singer's discretion. It is a purely musical effect and has, as a rule, no direct bearing upon the text other than a sympathetic one; he must choose for himself where he will employ it. It is chiefly associated with the return of the first subject, and is generally accompanied by a *ritenuto* followed by the *tempo primo* of that subject. In its *rubato*, therefore, it strictly conforms to the conditions spoken of above. The necessity for such strictness can be illustrated by the following two widely differing examples:

Mon   mal-heur - eux   a - mour. Bois é - pais

The first is from a song ("Bois épais" by Lully) which is sung *adagio* to begin with. If, after the additional slowing down contained in the *ritenuto*, the resumption of the *tempo primo* is not strict, the whole song becomes debauched.

The second ("Trottin' to the Fair," Old Irish, arranged by Stanford) is marked *allegretto* and derives its charm from its insistent rhythm (the trotting of the pony). Here the *rubato* is simply a "hold up" to render the rhythm more telling than ever when it returns. The strict return of *tempo primo* in the first was defensive, in the second offensive.

In purely strophic songs where the return of the subject is in the form of a "tag," reiterating the same sentiment in each verse, that tag is certainly stronger when not joined on by a carry-over. The sentiment seems to gain strength by isolation. In Schubert's "Litanei," even though the preceding phrase ends on the fourth and has a natural progressive tendency downward, the final "tag" is stronger and less sentimental if started afresh each time, and its importance as the motto of the song is emphasised.

In the *Rondeau* where the first subject recurs at least twice, the musical feeling seems to demand variation of treatment even of the same sentiment. Here the singer would do well to keep his carry-over till the *last* recurrence of the subject. "Plaisir d'amour" by Martini is a case in point. If in the second verse he breathes after the words "un autre amant," and starts his "plaisir d'amour" afresh, then the carry-over from the "changé pourtant" to the "plaisir d'amour" in the final verse will round off the song.

In Schumann's "Du Ring an meinem Finger" the carry-over at the second re-entry of the subject conveys a directly illustrative meaning. One can see the speaker's ecstatic gaze into space gradually return to the ring upon her finger, and can follow its actual trajectory in the carry-over. This is, of course, read into it by the singer. The composer has given no indication of it, but has trusted to the return of the *tempo primo* (which, as usual, he has not marked) for his effect.

The carry-over being essentially musical and *bel canto*, is dependent upon perfected technique. The singer cannot ride off on dramatic substitutes, for there are none. The carry-over, like all other *rubato* effects, has no inherent merits of its own, but depends upon the *tempo primo* for its livelihood. No singer should attempt to use it unless he has plenty of breath in reserve with which to tackle the *tempo primo* when he gets to it.

THE MELISMA

The *Melisma*, or quasi-florid *cadenza*, on the other
hand, is directly illustrative, either of the immediate
words or of the mood of the song.   It gives the singer
a purely vocal passage, in which by *rubato* and colour,
used *con discrezione*, he can illustrate the sentiment
without being trammelled by diction.   Probably the
greatest examples are in Bach.   It is hard to say whether

the passage from the *St. Matthew Passion*, or

its counterpart from the *St. John Passion*, is the more
beautiful, the more illustrative or the more moving ; while
for sheer ferocity the famous passage from the latter work,

taken with its context, surpasses in realism anything
of to-day. All three are extremely difficult vocally,
but, when sung by one who appreciates his privileges,
move the hearer profoundly. The second is probably
more difficult than the first, for it is longer and has a
rhythmic accompaniment, and is consequently less *ad
libitum*, but both depend upon pure phrasing handled
with emotional effect. They illustrate one of the most
touching and human moments in the world's greatest
story, and the singer should come to it well prepared.
The third is so vivid in its illustration that it dictates
its treatment. The rhythmical accents must fall like
lashes, callous, relentless, regular, with no variation
whatever. Here, if ever, the rhythm carries an accent
on every beat. The passage must be so physically
handled that it never pauses for a fraction of a beat, and,
in addition, the natural and illustrative *diminuendo* of
the descending phrase, and *crescendo* of the rising, must
be given graphically. This entails considerable reserve
power and dexterity of manipulation, and the singer
will do well in this, as in the other two, to memorise
the passage, so that he may have all his faculties free
for concentration upon its delivery.

The short *melisma* — so short as to be scarcely more
than a highly-developed turn — at the end of the verses
in Schubert's "Ungeduld" is simply illustrative of mood.

The sentiment, as the name implies, is "impatient,"
and the illustrative florid passage, being just an

expression of exuberant spirits, is rattled off with no appreciable *rubato;* whereas in Hubert Parry's "A Lover's Garland" the *melisma,* though the same in each verse, expresses a different sentiment, and is handled differently, each time it occurs.

I'm weav - - - ing too.

conveys the idea of weaving;

for He - lio - do - ra's brow.

the feeling of triumph at the completion of the wreath of flowers, and its offering to Heliodora; and

In-to her bosom, O hap - py they!

*Without taking breath, if possible.

the singer's sigh of admiration as he pictures it upon her brow and sees in imagination the petals fall into her bosom. (The composer is not responsible for these detailed expression marks; he left the interpretation to the singer.)

Again, in the same composer's delightful "Laird of Cockpen," it is used to illustrate in turn (1) the length of the lady's pedigree, (2) the egregious pomposity of the Laird, (3) her curtsey of refusal; and the singer consequently has splendid opportunities for illustrative tone-colour. Compare also his soprano song "The Maiden."

The *melisma* in the *Lied* or Song, being free in its handling, demands strict resumption of the *tempo primo* the moment it is over.

THE FINISH OF A SONG

A song can be marred — if not made — by its finish. All its points, if the song has been sung as a whole, seem to gather together for that climax, and here, if ever, the singer's sense of touch should be true to him.

As said before, you can tell a master of style by his cadences alone, and, as usual, those cadences are surrounded by enemies. The majority of songs end in a movement either down or up. The commonest way of emphasising a close is by the *rallentando* accompanied by the *portamento*. The *portamento*, called more commonly and appropriately the *slur*, is the bosom friend of sentimentalism, cheap effect, and contraltos. It is so dangerous, and has such a debauching effect upon sentiment when mishandled, that it is never allowed in pairs. This is not optional or a question of taste; the laws of singing do not allow two consecutive slurs. The contralto with her arm round the first, looks at the other with a longing eye, hesitates, catches sight of the policeman and passes along in tears. If it were suggested to her that she might be better parted from the first as well, she would prefer suicide. Though she does not believe it, the end of her beloved "Caro mio ben" would be infinitely stronger and cleaner without it, and more effective in the end. The slur, no doubt, does convey the physical impression of approaching climax or the end of a song, and where the style of the music or the innate sense of the text demands that outward form of expression it is both effective and legitimate; but it has been exalted out of its proper sphere to such a giddy position that it has lost both its head and its manners. There is an innate vulgarity about it which will out whenever it is given a free hand. It is hard to say which is the more dangerous, the downward or the upward slur — to the contralto probably the downward, as giving more scope for fog-horn effects. The ear has provided for both to some extent by exacting the touching, or anticipatory sounding, of the final note as a toll to clean phrasing; but both have the disadvantage, when occurring between the penultimate and final beats, of giving an inflated value to the unimportant note, word, or syllable.

The slur is the ally of small phrasing. To the salesman who has been dealing out snippets, the "workoff" of a roll of shoddy at the day's end is big business. The broad phraser, on the other hand, has plenty in reserve and, from the very size of the whole, can afford, if he will, to end even in small. Would any master of style end the "Erlkönig" with a slur between the words "war todt," or even treat them as a climax? Would he defile the simplicity of Franz's intimate little masterpiece "Im Rhein im heiligen Strome"

N

with a slur on the final "genau"? Would he slur from "till" to "I die" in the last line of the old Irish "Gentle Maiden"; or would he dare to slur down the last "Ich grolle nicht"? If the phrasing in each case were big it would give the slur the lie.

There are cases, on the other hand, where the slur is deeply expressive of the meaning; such as the end of Schubert's "Doppelgänger,"·where the slur, almost dragged down

gives the picture of the man's head sinking into his hands, and links up the tragedy of to-day with that of old times; or the final words of the "Dichterliebe" — "Schmerz hinein" — where, after a *Kunst-pause* between the two words — to sum up all that has gone before — a definite slur takes all the singer's love and suffering, lays them in the great coffin, and sinks coffin and all *down* in the depths of the Rhine.

There is a species of half-slur of which the last two notes of Schumann's "Widmung" furnish an example. The *semi-portamento* from the *res* of "bess'res" to the "ich" is a *quasi-abandon* to round off the *abandon* of the whole song and its sentiment, and as such is forgivable. But nine slurs out of ten are rank sentimentality.

There are various legitimate helps to "ending" effects; such as the *Kunst-pause*, or "artistic" pause, referred to above; the emphasising of the initial consonant or aspirate, as the case may be, of the final word; and the appropriate use of the anti-climax.

An excellent example of the first two is the ending of Schumann's "Du bist wie eine Blume."

So rein und schön und *hold.*

A slur from the leading-note to the tonic would be detestable, and, from the value allotted to the former by the composer, unjustifiable. Not only would it give an undeserved prominence to the conjunction *und*, but it would discount the majesty of the word *hold* ("noble") by tacking it on to its insignificant subject — a very disreputable morganatic marriage. It also discounts the strength of the aspirate in *hold*, which is vital to the nobility of the word. Unquestionably the way to end the song is by making a slight "Kunst-pause" (infinitesimal in duration, but definite) after *und*, and by slightly accentuating the following aspirate — both words sung *pp* and with absolute simplicity.

(The leading-note from its nature is generally on an unimportant beat and equally unimportant word or syllable, and does not deserve slurs or prominence of any sort, except in a rising *crescendo* leading to a *forte*, when its inflated value is lost in the strength of its resolution on the tonic.)

The *anti-climax*, in its right place can be the most effective of climaxes. Its effect, being unexpected, gives a thrill which the obvious can never give. The only thing to remember about it is that such effects must be true; the dragged-in anti-climax is an abomination. The last words of the "Erlkönig" — *war todt* —

are a case in point. One could not make a climax out
of them if one would. The *Sturm und Drang* which
have gone before are far bigger dynamically than they.
To make a physical climax of them, the singer would
have to shout. But the father did not shout; he was
stunned. He did not even speak; he *saw*. The more
quietly the narrator tells you that the boy was dead,
the more dramatic his climax.

Take, again, the last words of "Allnächtlich im
Traume" from the "Dichterliebe."

Here, superficially, the despair of his awakening would
imply an outburst on these very words; but the outburst is begun at "ich wache auf," and is over at the
word "Wort." Then comes a rather long *Kunst-pause*,
in which his head drops into his hands while he dumbly
racks his memory, and then, quite quietly, the plain
statement "hab' ich vergessen," more full of hopelessness than all that has gone before.

In many songs such as these, anti-climaxes are
another name for simplicity. The mere statement of
a fact, the simple answer to a question, the expression
of a quiet thought are the strongest in their effect.
The last words of Walford Davies's "When Childher
plays," and the answers to the questions throughout

Harold Darke's "Uphill," are models of the inherent
strength of simpleness. But the greatest anti-climax
in musical literature is probably the famous short
recitative :

from the *St. John Passion*. What drama or pathos
could the singer put into those words that are not
inherent in them already? Not only do they immediately follow one of the most touching airs that even
Bach ever wrote, but they are in themselves the end
of all things, the end of the greatest tragedy in the
world. No human interpretation could add one jot

or tittle to their touching simplicity. One feels that
with the "Man of Sorrows, despised and rejected of
men," such a simplicity is in keeping; one leaves them
unadorned, and says them, and no more. And yet if
such an abstinence were forced, or sounded untrue, the
narrator would do better to sing them with all the
emotion of which he is capable, than risk any
approach to artificiality.

In the ending of most songs the governing element
is *pace*. If the singer has absorbed his atmosphere,
kept up the march, and sung mentally through his
rests, then atmosphere, march and rests must be consistent to the end; otherwise he will not have treated
the song as a whole. Because the approach to home
favours a slower stride that stride need not degenerate
into a lounge. The end of a song, like its interpolated
rubatos, grows weaker the farther it strays from its
fundamental rhythm. The closer the singer holds,
consistently with his climaxes, to that rhythm *to the
end*, the stronger that end and the cleaner his style.
The broader the treatment of his song, the less fuss he
requires to end it. The last phrase of the "little"
song is generally a Gulliver among the Lilliputians.

CONSISTENCY

This question of consistency is all-important to the
song in large. The temptation to drop a song temporarily from its level in order to illustrate a word or a
sentence is overpowering. But the gain to the words
is a poor compensation for the loss to the music; and
the music's the thing. Instances could be quoted
without number, but one will do:

mf   cres. poco a poco al fine   • • •

Wollst end-lich son-der Gräm - en aus die - ser Welt und neh - men durch ei - nen sanf - ten Tod; und wenn du uns ge - nom - men, lass uns in Him - mel kom - men, du un - ser Herr und un - ser Gott!

This is the beginning of the last verse of Schulz's (1790) famous "Abendlied." The two previous verses have been sung *piano*, and devotionally, like a vesper hymn. In the last verse — where the hymn is treated as a song — the singer works it up emotionally by a steady *crescendo* from the beginning to the end. If, in answer to the call for appropriate dynamic and colour treatment of the words "einen sanften Tod" ("a gentle death"), he drops from his level to illustrate them, his *crescendo* is no more, his consistency is gone and his song become a thing of shreds and patches. Such hyper-conscientious illustration, like its twin-sister over-elaboration, is fatal to interpretation, and wherever it shows its seductive charms must be spurned with the heroism of a St. Senanus.

(Compare Hubert Parry's "When lovers meet again," in which the words "Peace rocks the world in calm" are marked *crescendo*.)

WORD-ILLUSTRATION

(*N.B.*—This should be read in connection with Main Rule III.)

MANY readers will remember the famous onomato-pœic πολυφλοίσβοιο θαλάσσης ("the loud-roaring sea") of their schooldays. Its dynamic potentialities can be expressed in actual musical signs as follows :

$$mf \xleftarrow{\hspace{2cm}} ff \xrightarrow{\hspace{2cm}} pp$$

πολυφλοίσβοιο θαλάσσης

The very eye can see the rise and fall of its wave. But it has more in it than that. Its polysyllables tumble over one another to build its mighty billow ; on the β it bursts with a boom ; roars on the οι high up the strand ; then hisses back over the sibilant shingle of θαλάσσης.

If a couple of words can vividly convey all this, poetry and music together should be worthy of their hire.

Rule III. dealt with Speech in Song, and concerned itself mainly with the Vowel. Word-illustration deals with Song in Speech, and has to do chiefly with the Consonant. The first was passive, the second active. One demanded no more than that the pure vowel of "love" should be left in peace ; the other asks that the aspirate of "*h*ate" shall be accentuated for purposes of dramatic illustration.

With the vowel the singer builds his structure in the large; with the consonant he rounds off the edges and adds the ornamentation. It follows, therefore, that consonantal illustration is confined to the shorter note-values. Both belong to the twin-sisters "Voice and Verse," and work together in harmony; and the vowel in this branch of the art gladly accepts the precedence of the consonant and backs it up — with colour and pressure-values. By taking advantage of this co-operation apparently ordinary words, which would otherwise depend upon their musical setting or dynamic sound for their sole effect, achieve an active life of their own.

An accentuation of the initial consonant or consonants, or aspirate of the following words, reinforced by a "push" on the succeeding vowel, promotes them from mere vehicles of sound to co-operative illustration:

> strong, crude, frost, bright, curse, blow, hate,
> blood, ring, shame, stunned, laugh, etc.

Others gain pictorially by a distinct *crescendo* and *diminuendo* on the vowel following the accentuated consonant:

> broad, creed, proud, tall, shine, hate (where the
> sentiment is lasting or deep), freeze, queen,
> deep, cool, calm, moan, mourn, scythe (to con-
> vey the idea of its swath), gaze, swoon (ex-
> pressing gradual unconsciousness), waft, closed,
> etc.

Many of these gain enhanced effect by a distinct
sounding of the final consonant or consonants as well:

hate, stunned, shame, shine, round, swoon, gaze,
dream, peace, flame, blush, fresh, freeze, sing,
drone.
."Still his hurdy-gurdy drones and drones
away."

The droned *n* here is far more expressive than any
holding of the vowel sound.

Some monosyllabic words are sung as though they
were followed by an exclamation mark (!) :

struck ! pluck ! fife ! wept ! haste ! dip !
drip ! shock ! shook ! dance ! swift ! quick !
snap ! kissed ! touch ! gnash ! etc.

The same accentuation of initial consonants applies
here too; the finals, however, are shortened with a
distinct "snap."

Dissyllables have the double advantage of the use
of the consonant and the contrasts between the long
and the short foot in the word. A strong push, or
gentle blow (as the case may be), on the long foot
(accompanied by accentuation of the consonants), followed by a distinct shortening of the short foot, gives
a pictorial value to such trochees as :

slender, stately (as though the speaker acknow-

ledged a curtsey) curtsey, dreadful, ghostly,

worship, purple, frozen, etc.

These are all *pushes* and, with the exception of
the *z* in the last word, the secondary consonants,

though distinctly sounded, are of secondary importance:

$$\text{láughter, túmbled, pómpous, báttle, ghástly,}$$

$$\text{búgle, thúnder, péaling, trémble, rápture,}$$

$$\text{trámple, véngeance, trúmpet, etc.}$$

These are *blows*, and the initial and secondary consonants require greater emphasis than in the *pushes*.

Many trochaic dissyllables too imply an exclamation
mark:

> vivid ! chatter ! scatter ! hurry ! rattle !
> riddled ! sudden ! happy ! terror ! little !
> bloody ! sunny ! etc.

and should be generally sung as though it were there.

Some trochaic words, either from long use in singing,
or from a natural leaning to pictorial effect, have become
almost spondaic in accentuation:

> lilies, rosebud, hollow, silent, singing, satin,
> vision, etc.

The accent is, of course, undeniably on the first syllable, but the second has acquired a false value in
excess of the ordinary second foot of the trochee. The
word *vivid* is far more vivid if both syllables are
given exactly even values; the trochaic value would
actually detract from the illustrative powers of the
word. Likewise *shudder*. *Sudden* is a peculiar word.
As an adjective it is far more effective when equal
values are given to both syllables. The sound carries
the *rat-tat!* of a double knock. But in the adverb
suddenly the whole weight is thrown on the first foot

of the dactyl, the two short feet being flicked off as almost of no account. The same applies in less degree to *passion* and *passionate*.

Certain trochees are given an inverted time-value in notation either by composer or singer, as the case may be. Thus *echo*, which as a trochee would naturally be written ♩. ♪ or ♩ ♩, is, and should be,

   e - cho   e - cho

generally sung, if not written ♪ ♩..

   e - cho

The same applies to

> horror, pity, lily, petal, chatter, meadow,
> yellow, withered, dewy, honour, battle,
> precious, any, bitter, nothing, etc.,

and a good many of the exclamatory trochees. Such a readjustment of values has the natural effect of still further accentuating the primary accent of the word. Its effect upon *never* and *ever* is remarkable.

Iambics, *mutatis mutandis*, work on the same principle. Here the *final* consonants strengthen, and are strengthened by, accentuation :

$$\smile \; - \quad \smile \; - \quad \smile \; - \quad \smile \; - \quad \smile \; - \quad \smile \; -$$
resoun*d*, remor*s*e, begi*n*, conse*nt*, posse*ss*, devou*r*,

$$\smile \qquad -$$
blasphe*m*e, etc.

The singer must ever be on his guard against the psuedo-Italianism referred to earlier, and keep clear

$$\smile \; - \qquad \smile \; -$$
of resoun*d*-a, begi*n*-na, or, where a vowel-sound ends

$$\smile \; - \qquad \smile \; -$$
the word, obey-ur, renew-ur, etc.

Polysyllables sing themselves. No singer with a

feeling for beauty of sound could resist the call of such dactyls as

exquisite, beautiful, wilderness, wonderful, etc.,

or of such words as

delightful, surpassing, tremendous, colossal, beloved, contented, honeysuckle, inscrutable, etc.

Many of the dactylic flower-names have optional, almost equi-syllabic, values:

dáffódíl, jéssámíne, víólét, cólumbíne, hýácínth

resembling here such words as

sycamore, sinister, cavernous, shivering.

The word *beāutĭfŭl* is a pure dactyl, and though often written on three even notes, should never be sung except as a dactyl. The singer, no matter what the written notes may be, should so manipulate the word as to restore it to its proper values. Such "optional" variations of scansion are generally concessions to musical notations, and are sparingly used by the composer who has had a classical education. Their pictorial effectiveness is small.

Some words by the prolongation or manipulation of their consonants can actually convey a sense of the *time-duration* of the sentiment or action.

"watc*h*ed." By holding on the *ch*, the singer can actually dramatically imply a prolonged gaze or vigil; such a prolongation would be out of place in the word *watchman*, unless he happened to have his eye on a burglar.

"clo*s*ed." The door can be shut slowly or quickly at will by the length given to both vowel and consonant.

"hush." The *sh* can by its duration express either a command for silence or the croon of a cradle song.

"vanish." By means of the *sh* the ghost can either disappear in a flash of lightning (vanished !) or fade slowly out of sight.

"gone" can also with both vowel and consonant convey either sudden or gradual disappearance.

"touched." The *ch* may literally "touch" with an exclamation mark, or, by lengthening, smooth a fevered brow.

The most actively pictorial consonants are *n* (drone), *m* (hymn), *s* (whisper) ; and the strongest combinations *m* or *n* followed by another consonant (singing, clank, clanging, tinkle, trumpet, tremble, and that most beautiful of words — remember), *ch* (watch), *sh* (shudder !), *ss* (restless !).

With these the actual physical sound of the word can be conveyed vividly without sounding ridiculous.

It must ever be remembered that a song, however dramatic it may be, can only be a reproduction *en miniature* of the scene or sentiment, and illustration must be drawn to scale. It must also be remembered that whenever the consonant, as above, assumes command, its extra values are taken at the *expense of the vowel*. There must be no lengthening of the note or bar to compensate the vowel. She is not very much to be pitied ; she has a good time of it as a rule, and the high note at the end of the British ballad will always give her something to live for.

Word-illustration to the British contralto is generally confined to the word "struck !" Some famous contralto of bygone days once *struck* a chord on the organ "like the sound of a great Amen," and on the snowball principle Anglo-Saxon contraltos have thus *struck!* it ever since. It does not matter that such a lightning-like *martellato* could never possibly reproduce any actual resemblance to an "Amen"; she has made her concession to drama, and can return to "Lascia ch' io pianga" with honour satisfied.

An example of subtle word-illustration is contained in the following passage from Arthur Somervell's "She came to the Village Church" ("Maud" cycle, No. 3):

And once, but once, she lifted her eyes, And suddenly, sweetly,

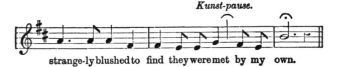

strange-ly blushed to find they were met by my own.

"once." The sounding of the sibilant in some un-explainable way emphasises the simplicity of the lifting of her eyes. They had been on her book throughout the service until then.

"suddenly." This must be a pure dactyl. Not only should the two short syllables be barely sounded, but the vowel *u* should be so lengthened (sung *pp* with a very white "wondering" colour) that it should *not* sound sudden. The colour did not *fly* to her face; it *stole* there.

"sweetly." The sibilant and the *crescendo* and *diminuendo* on the *ee* (a very pure *ee*) are expressive of the word.

"stranğely̆." A pure trochee with the short syllable hardly sounded. The sibilant and vowel must have the same value and colour as in "suddenly" to convey *wonder*.

"blu*sh*ed." The colour *stole* to her face slowly. A lengthening of the *sh* can tell you so.

The "Kunst-pause" after "met" implies a question. "By what?" Answer, "By my own."

"By my own" must be sung with extreme simplicity. It is a climax by anti-climax, far more powerful than any laboured air of triumph.

The *carry-over* from "eyes" to "and suddenly" shows the concentration of his gaze. It almost implies a gesture as though when she lifted her eyes he unconsciously half rose from his knees. Strange to say, the "suddenly" sounds more unexpected (when sung with appropriate tone-colour) joined on than separated.

Here we have in six bars of music not only five subtle examples of word-illustration, but a *Kunst-pause*, a *carry-over* and an *anti-climax* into the bargain!

To some people language has no pictorial value; to them it is but a necessary means to commonplace ends. But even the slovens have their apologists. In every concert-room to-day, where sweet music is bedraggled in the telling, one can hear the deprecatory remark that English is a poor language to sing — made by the possessors of the Elizabethans, of the "Blest Pair of Sirens," of the "sounding brass and the tinkling cymbál," of the 104th Psalm and the Book of Job!

DISREGARD OF WRITTEN EXPRESSION MARKS

At first blush this looks like rank heresy. But were expression marks made for man or man for expression marks? One thing is certain, the greater the music intrinsically, the fewer the arbitrary expression marks. Bach and Schubert were not afraid. It is only the cheap-jack who overloads his score with our old friends cheap effect and over-elaboration. The composer's expression marks are the expression marks of the individual composer and may not suit the individual singer. The strong composer knows this well, *and if his music is strong enough to stand as music,* hands over its interpretation to the individual, conscious that, if that individual is to be trusted, the great essentials will be treated with profound respect. Many a modern song the interpreter looks at with a shudder. Riddled with expression marks and even breathing marks, hedged in with arbitrary directions, radiating polyglot colloquialisms, it looks like a barbed wire entanglement. Singer and accompanist smile at one another, study the song as a whole, and sing it their own way.

If this book preaches anything, it preaches individualism. Communism in music is another name for conventionality and reeks of stale beer. The skilled labourer is worthy of his hire, and the hand-made will ever be greater than the machine-made as a work of art. To the strong composer the strong interpreter is a man to be trusted. He knows that all his *pianos* and *fortes,* and *affrettandos* and *con grazias,* while necessary for the imitator, are superfluous or restricting to the individual, and he leaves the individual

o

untrammelled in their handling.   A touch of enterprise is worth all the expression marks in his vocabulary.

One singer has different gifts, vocal and temperamental, from another; the rigid following of expression marks may spell success for this and failure for the next.   But there is a broader aspect to it than this. A song may affect one singer emotionally in a quite different way from another.   The actual sound of the music, quite apart from the singer's gifts or limitations, may move one man to tears and another to anger, and another even to laughter.   Brahms's "Vergebliches Ständchen" is treated by some singers in a spirit of semi-seriousness and coloured accordingly.   Others it does not affect in the same way.   To them it is a little cinematograph scene with subdued laughter (on one side at any rate) running all through it, played at a great pace and over in a minute.   To ask either arbitrarily to adopt the other reading is to ask them to swallow their convictions.

In his treatment of the well-known masterpieces, the singer is faced by Tradition.   Tradition has its advantages and its drawbacks.   There are some traditions which no individualist will ever transgress. These are built on such solid foundations, and the structure is so beautiful architecturally, that to touch them would be vandalism.   Some traditions again — in matters of *tempo* especially— have been handed down by the great interpreters and should not be departed from — for change for change's sake is fatuous.   But tradition can be exalted into a fetish and may be just as futile in practice and dangerous to trust to in emergency. Traditions in singing have generally been established by some great singer.   That singer has had vocal and mental gifts far above his fellows, and the standard of his traditions may be as far above other men's heads after his death as his interpretation was in his lifetime. Because Stockhausen sang a Schubert song like no one else in his individual way, it does not follow that every singing pupil of every Conservatoire should be fitted physically and intellectually to adopt the reading of the greatest individual master of Lieder-singing of our time. Tradition has been responsible for the smothering of many a talent, if not for actual physical damage to the singer. Tradition of the great interpretations is a thing of beauty and a joy for ever. It should stand on a pedestal for all to see and compare with; but it should never be made a table of commandments. Individuality and imagination are the greatest gifts of all; useless tradition cramps the one and kills the other. But individuality must be as responsible as imagination is spontaneous; the singer who alters a composer's marks for originality's sake, or insults tradition from wantonness, is an impostor and a foe to progress.

"*Con discrezione* in that and many other things."

In conclusion, the interpreter should bear the following axioms in mind:

Don't make too many climaxes in your song; they spoil one another.

Don't hammer in your nails with a sledge-hammer; the music does it for you.

Don't forget that it requires as much trouble to sing "Die Rose die Lilie" as "Der Erlkönig," and that each is as perfect a work of art as the other.

Don't forget that in most cases increase of effect is gained by broadening and slowing down rather than by quickening, provided always that such broadening is in keeping with the fundamental rhythm; it is only in the quick song, as a rule, that climax is achieved by increase of pace.

Don't forget that time-signatures imply a promise which must not be broken.

Don't forget that rhythm has to be indemnified for the use of *rubato*.

Don't paint your details at the expense of your picture.

Don't forget that long phrasing has no intrinsic merits of its own. It is great as an enhancement of rhythm and the friend of the song in large, and therefore of style; when it does not fit with these, it carries no weight.

Don't "interpret" where interpretation is not wanted.

Don't point out the joke in humorous songs.

Don't forget that the music's the thing.

To the Accompanist.

Don't make a *rallentando* in the opening symphony immediately before the entrance of the voice, unless it is either specially marked in the score or agreed upon by both performers as advisable, and rehearsed. When unexpected it is liable to throw the singer who is observing Main Rule II. out of his stride. This applies particularly to rhythmical songs.

Don't forget that it is easier for the singer to hold you back than to hurry you on.

Don't make "pianistic" effects that have not been rehearsed.

Don't wait for the singer, except as part of the rehearsed interpretation — if you can trust him.

When all is said and done, there are many things that can never be put in black and white; things that come from nowhere and make the singer change his whole reading even at the last moment on the very platform. Such are little private messengers from his brain to his tongue, intimate little spirits who whisper in his ear at the very last, or draw aside the curtain for an instant and show him a glimpse of fairyland, or lure him on to break the law or invent new laws for the nonce. He may trust them. They are the ambassadors of temperament, the little call-boys of imagination, and where they call him he may safely follow. The true interpreter will break every rule and conform to none of these standards when it suits him so to do. If he is a man of refinement, he will never break the unwritten law; if he is a man of honour, he will not betray the composer. But one thrill is worth a thousand orthodoxies.

PART IV

THE CLASSIFICATION OF SONGS

IN most branches of music England in the last forty years has advanced with giant strides; in one she has stood still, if not actually gone back — England's "popular" song is utterly unworthy of her. Its "popularity" and its unworthiness are due to the working of a certain commercial system, which there is no intention to discuss here further than to say that the system in itself has great possibilities, and if applied to the advancement of good instead of bad song would be of undoubted service to music. As it is, it must be confessed that the "popular" song of to-day in healthiness of sentiment and workmanship shows a marked decadence since the days of Hatton or even Virginia Gabriel.

The song makes by far the widest musical appeal. It is bought and sung by thousands who have never heard an orchestral or chamber concert in their lives. Its power for good or evil is proportionately great; and there is no doubt that the low standard of the "popular" song by its very universality bars progress in what is probably the most musically-gifted country of to-day. So long as the public are satisfied with bad stuff, so long they will be given it. The change must come from them. They think that they are supplied with what they demand; in reality they are demanding what is supplied. If they will but realise that the average "popular" song at the average "popular" concert is not sung on its merits, but for commercial reasons, they may begin to see the light. The Englishman has a natural and rational prejudice against the charlatan; let him once understand that the singing of a song in public by the professional singer is no guarantee of its intrinsic virtue, and the days of the bad song will be numbered. The step from realisation to discrimination will be a short one, and the barrier across the people's right-of-way will be demolished for good and all. The "cheap" song is the property of the cheap-jack composer, and it begets a line of slovens. So long as it holds the field, there will be no room for interpretation.

Songs may be roughly classified as Folk-songs, Simple Strophic Songs, and Art-songs.

Folk-songs. These are practically always in strophic or stanza form.

Simple Strophic Songs. These are modelled on the idea of the folk-song and provided by the composer with an instrumental accompaniment. They are the polished and conscious editions by artistic composers of what in the folk-song was the result of unconscious, purely human, instinct — trying to explain itself and eventually doing so by means of melody.

Art-songs. In these the music is not bound by the stanza form of poetry, but forms a running commentary on its contents with a pictorial and emotional background provided by the accompaniment.[1]

There is no particular reason why songs should be gathered into groups except for convenience' sake. If they can be classified, the same main idea of interpretation, or, rather, the same attitude of mind, or mood, towards the individual song, will practically apply to all those which come under the same heading. Many songs belong to more than one group, but even in these one group indicates the mood while the others refer to secondary considerations. The classification of songs is governed by the question — What dominates the course or constitutes the atmosphere of the song *as a whole?*

In some songs the answer is written so that he who runs may read. They carry their meaning so openly on the surface that their interpretation is a merely physical matter. This applies, as said before, to the *bel canto*, florid and (in a less degree) rhythmical groups, and even to those songs of reminiscence in which "the days that are no more" are actually and verbally made the subject of the poem and its musical setting.

But the most interesting songs are psychological, and on the sensitiveness of the singer depends the individuality of his reading. One thing is sure — if he is individual and sensitive he will make that reading his own for good and all, and will let everything go that interferes with it. The song *as a whole* will so take possession of his mind that note-values and expression marks, and all other details which do not come into line with his imagination, will go down

[1] For the above admirable definitions the author is indebted to a lecture by Mr. Walter Ford.

before the reading in large. But like most heretics he has to start by being sincere.

In all essays upon attitudes of mind it is impossible to escape a certain air of priggishness which is even more annoying to the reader than the didacticism of rules; but the sincerity of the heretic must be pleaded in mitigation of damages. To say, for instance, that in *Atmospheric* songs the voice-part is invariably subjective, that the greatest tribute the audience can pay a singer is to be only subconsciously aware of his share in the song, that the song, *as a whole*, if the singer has played fair, should have so absorbed the mind of the listener, that the singer's personality has been temporarily forgotten, sounds highfalutin', but is none the less true. To say, too, that the *mental* treatments of *Contemplative* songs and *Songs of Address* are so strongly opposed, that the singer must feel his *physical* voice to sound in the one *introspective* and in the other *telepathic*, may sound a mixture of jargon, but it genuinely illustrates the intimate correlation of the mental and physical in the absorbing and interpreting of one song and another.

It may be said of all songs that the more subtle the atmosphere and resultant mood, the more interesting the study of interpretation. The obvious always inclines towards the merely physical.

Songs may be roughly classified as:

1. Atmospheric.
2. Dramatic.
3. Narrative.
4. Songs of Characterisation.
5. Songs of Reminiscence.
6. Contemplative.
7. Songs of Address or Ode Songs.

8. *Bel canto,* Florid, and Rhythmical.
9. Ghost Songs.
10. Songs of Question and Answer.
11. Humorous and Quasi-humorous.
12. Folk-songs.

(1) *Atmospheric Songs,* such as:

Der Leiermann. } Gretchen am Spinnrade. }	Schubert.
Feldeinsamkeit. } Auf dem Kirchhofe. }	Brahms.
The Fairy Lough.	Stanford.
Nightfall in Winter.	Parry.
A Widow Bird.	Luard Selby.

and many of the most beautiful of Debussy's songs, and of the modern school in general.

The word "atmospheric" does not necessarily imply anything to do with the weather, or that other songs have not atmosphere, but that these depend upon atmosphere as their dominating characteristic. Many, such as the two Schubert and the Luard Selby songs quoted above, have a definite insistent rhythmical accompaniment, which absorbs the attention and automatically renders the voice-part subjective. Atmospheric songs are rare, but by far the most subtle and interesting of all song-literature.

(2) *Dramatic Songs,* such as:

Der Erlkönig.	Schubert.
Waldesgespräch.	Schumann.
Vergebliches Ständchen.	Brahms.
Ethiopia saluting the Colours.	Charles Wood.
La belle Dame sans Merci.	Stanford.
The twa Sisters o' Binnorie.	Traditional.

in which the scene is acted in song and the characters differentiated by tone-colour. The last two are, of course, ballads, but so dramatic in treatment as to merge their narrative qualities in the action. "Vergebliches Ständchen" is, as stated before, a little comedy played by two characters.

(3) *Narrative Songs*, such as:

Die beiden Grenadiere.	Schumann.
Die Forelle.	Schubert.

and the famous ballads of Loewe, and all ballads in which the *story* is the principal feature of the song. Many dramatic songs, such as "Der Erlkönig" and "Ethiopia saluting the Colours," assigned to the last group, overlap with *Narrative* songs. The last named, owing to its symbolical significance, almost overlaps with *Atmospheric* songs.

(4) *Songs of Characterisation*, in which the singer assumes a character for the time being, and adopts its sentiments nominally as his or her own throughout. This applies to nearly all song-cycles, such as:

Die schöne Müllerin.	Schubert.
Dichterliebe.	
Frauenliebe und Leben. }	Schumann.
Maud.	Arthur Somervell.
An Irish Idyll.	Stanford.

or such songs as:

Der Rattenfänger.	Hugo Wolf.
Der Wanderer.	Schubert.
The Vagabond.	Vaughan Williams.

These songs, and *Narrative* and *Dramatic* songs, are all closely associated.

(5) *Songs of Reminiscence:*

Der Doppelgänger.	Schubert.
Es blinkt der Thau.	Rubinstein.
Silent Noon.	Vaughan Williams.
When Childher plays.	Walford Davies.

and the many settings of "Tears, Idle Tears" and other poems which deal with "the days that are no more." A vast number of songs belong to this group. It commends itself naturally to emotional composition and singing. Practically all songs written in the historic present — as moving in song as it is detestable in literature — are songs of reminiscence, and consequently extremely emotional. The scene described is pictured *in the memory* of the singer. "Feldeinsamkeit," written in the present tense, is purely atmospheric — just a dream-picture of blue skies and summer fields and drowsy happiness; but in Vaughan Williams' setting of Rossetti's "Silent Noon" not only have we the blue skies and green fields and the dragon-flies and the hot summer day, but the whole song throbs with the love of a man for a woman on that day *in the past, told* us to-day.

So in "Es blinkt der Thau" the very words *dass es ewig so bliebe* make us feel in our hearts that the day was long ago, and the speaker is living it through again in memory. The "days that are no more" must not be *acted.* The picture is too intimate, too sacred, to drag before the footlights. It is gone and dead and buried, but its memory is as green to-day as though it had *ewig so geblieben.* "Der Doppelgänger" will be discussed later. The word "remember" is never necessary to the song of reminiscence.

(6) *Contemplative Songs.* Another large field for the composer.

In der Fremde (No. 1). ("Aus der Heimath.")	Schumann.
Die Mainacht. Todessehnen.	Brahms.
Auf dem Wasser zu singen. Gesänge des Harfners.	Schubert.
Plaisir d'Amour.	Martini.
When Childher plays.	Walford Davies.
Corrymeela. Cushendall.	Stanford.

Such songs are mostly introspective. They are the musings of the poet's mind expressed in music; they may be inspired by reminiscence, as in "When Childher plays"; or by things seen conducive to reminiscence, such as the summer lightning in "In der Fremde"; or by the state of the poet's mind, as in "Die Mainacht"; or by the mental vision of some individual or place — "Plaisir d'Amour," "Corrymeela" and "Cushendall"; or they may be a frank homily on any given subject, such as the *solitude* of the "Harper's Songs."

To be *contemplative* does not necessarily mean to be slow in *tempo*. Both Ernest Walker's and Battison Haynes's setting of the old words "Hey! Nonny No," are vividness itself, yet both are contemplative; while Hubert Parry's "Love is a Bable" is as unquestionably contemplative as it is rough-and-tumble in its appropriate treatment of the words.

Many contemplative poems, such as "O loss of sight, of thee I most complain" from Milton's *Samson Agonistes* or Dekker's "Art thou poor? yet hast thou golden slumbers," are nominally addressed to some thing or condition of things or individual, but are in reality contemplative (in the first a contemplation on blindness, in the second on "sweet content"). In others the personality of a place may so overpower the speaker's consciousness that the contemplation becomes an ode. Such are Stanford's "Corrymeela" (an Irish Idyll), in which the longing for home lifts the man out of himself and projects his very being back to his glens of Antrim; and "Cushendall" (song-cycle *Cushendall*), where the thought of the little valley expands in its expression into a veritable worship. Such a worship of things temporal, when expressed in song (cf. Schumann's "Du bist wie eine Blume"), automatically divides this group from the next.

(7) *Songs of Address or Dedication, or "Ode" Songs,* such as:

An die Musik. An die Leyer. Die Allmacht.	Schubert.
Widmung.	Schumann.
Wie bist du, meine Königin.	Brahms.
To Anthea.	Hatton.
To Lucasta. To Althea.	Parry.
The Roadside Fire.	Vaughan Williams.

This is naturally another large group. The song need not necessarily be directly addressed to the individual; witness "Die Allmacht" which, though nominally a contemplation, is in reality an ode in praise of God.

Most of these "Songs of Address" must presumably be semi-contemplative, and delivered to space. Their direct application to the victim is only imaginable on the stage; even the most insatiable prima-donna would fidget in time under a four-or-five-page enumeration of her virtues when delivered in the Sahara or the privacy of her boudoir. They are thus both detached and *telepathic*. But the true *telepathic* song is of the type of "Now sleeps the crimson petal" by Roger Quilter, which carries the old troubadour idea of the singer's voice being meant primarily for someone else, and being mentally projected into that room within the castle. The songs of this group, however, generally carry their treatment on the surface.

(8) *Bel-canto Songs*, such as:

Caro mio ben.	Giordani.
Ombra mai fu.	Handel.
The Self-banished.	Dr. Blow.
Auf Flügeln des Gesanges.	Mendelssohn.

and a vast proportion of the old school; all of which are not songs of interpretation at all, but depend upon being sung "instrumentally" with beautiful tone, phrasing and diction, and the following of technical rules — songs with a perfectly patent atmosphere. They include nearly all devotional songs, such as:

Litanei. Ave Maria.	Schubert.
Todessehnsucht.	Bach.
Abendlied.	Schulz.
Floodes of Tears.	Traditional.

Under *bel canto* come also florid songs which have no particular aesthetic value of their own, such as Bishop's "Bid me discourse"; and the greater part of purely *Rhythmical Songs*. The latter are either:

(a) Positively illustrative of some rhythmical motion, such as the trotting of the horse in Schubert's "Abschied" or in the old Irish "Trottin' to the Fair," or of the hammers in Brahms's "Der Schmied," or of the "rocking" in boat-songs and lullabies.

Or are (b) negatively representative of some individual or mood in which there is no particular variety of sentiment or colour, the rhythm by its strength obviating monotony. Such are:

Già il Sole dal Gange.	Aless. Scarlatti.
Dithyrambe.	
Das Lied im Grünen.	
Wohin?	Schubert.
Das Wandern.	
Corinna's going a maying.	Ernest Walker.'
The Old Superb (Sea Songs, No. 5).	Stanford.
The Vagabond.	Vaughan Williams.
Der Rattenfänger.	Hugo Wolf.

Some of these belong also to other groups, but one and all are dependent for their effect on unflinching rhythm. To apply the word "monotonous" to any one of them, even hypothetically, seems ridiculous; but rhythm was the means deliberately used by the composer for the artistic end, and the song if sung without it will collapse.

Or (c) use rhythm augmentatively for enhancing emotional or dramatic effect, as in Schumann's "Ich grolle nicht," where the insistent rhythm beats in the emotion of the voice; or Schubert's "Ungeduld," in which the "hurry" is conveyed by the rhythmical figure, not by interpolated *accelerando;* or in the "Gruppe aus dem Tartarus" (Schubert), where the relentlessness of punishment is hammered relentlessly

and rhythmically into the mind; or the same sentiment in "Dead, long dead," from Somervell's *Maud*
cycle, where the rhythm of *my heart is a handful of
dust* pounds like a pulse in the disordered brain.

All rhythmical songs make their effect by direct
appeal to the senses. Let the accompanist keep the
song ever "pushing on."

$\frac{6}{8}$ time (and in a lesser degree its kindred $\frac{9}{8}$ and
$\frac{12}{8}$) has remarkable rhythmical characteristics. One
and the same time-figure can by treatment be made
to illustrate a dozen things from remorselessness or
fresh air to dancing or cradle-rocking.

Compare the tremendous accents of Schubert's "An
Schwager Kronos" or "Dithyrambe" with the ecstasy
of Schumann's "An meinen Herzen, an meiner Brust,"
or with the swaying of the tree in "Der Nussbaum,"
or again with the pastoral beauty ($\frac{12}{8}$) of Bach's
"Beglückte Heerde"; or the thunder and fresh air of
Stanford's "Song of the Sou-Wester" (*Songs of the
Fleet*, No. 2), or "How does the wind blow?" (*Cushendall*, No. 6) with the gentle rocking of the wave in
his "Boat Song" ($\frac{9}{8}$) or "Fairy Lough" ($\frac{9}{8}$), or with
the lullabies of any period — nearly all in $\frac{6}{8}$ or $\frac{9}{8}$ time.

But it is the $\frac{6}{8}$ with the dotted quaver which give
the singer his opportunity. That dot has extraordinary
potentialities for good or evil, and the wise composer,
knowing its dangers, however he may use it in his
accompaniment, writes it sparingly for the voice. The
singer, rightly or wrongly, dots the quaver in $\frac{6}{8}$ time
where he likes. If he does it knowingly, with alternating even-quaver bars or beats, it is not only
legitimate but admirable; if he does it merely to help
himself along, he can turn a lullaby into a jig. Thus

P

in the beautiful old Irish melody, "The Gentle Maiden,"
the following passage,

There's one that is pure as an   an - gel, and fair as the flowr's of May.

if sung with strong accents and hyper-dotted quavers
would sound like an ordinary pretty dance-tune;
whereas by smoothing-over and manipulation of the
dot it can be promoted into one of the purest pieces
of *cantilena* in existence. The quavers here in the
first bar must be so little dotted as to be scarcely
distinguishable from even notes, and the dot must be
actually removed in the second bar; then when it
comes in the third bar it swings the melody on with
a gentle lilt all its own.

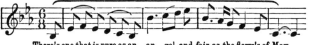

There's one that is pure as an   an - gel, and fair as the flowr's of May.

In nothing is the singer's sense of touch and natural
style more conspicuous than in the handling of $\frac{6}{8}$ time
in the voice-part as such and in its relation to the
accompaniment. The dotting or undotting of the
quaver is never noticed in detail by an audience, but
its use or abuse is vaguely felt and put to the credit
or debit of style unconsciously.

This *bel canto* rhythmical group includes also the
greater part of folk-songs and all the old German school
of Minnelieder, such as "Von edler Art," "All mein
Gedanken," "Ein schön' Tageweis'," "Wächterlied,"etc.

The interest of all songs belonging to this group is

musical rather than interpretative and their singing proportionately more difficult. (Basses are strongly recommended to study Purcell's " Ye twice ten hundred deities," which probably contains in a short space more varieties of *bel canto*, recitative, and styles of technique, than any song in existence.)

(9) *Ghost Songs*, in which the idea of the actual ghost or the supernatural is conveyed by the tone-colour. Such as:

Die Lorelei.	Schumann.
Through the Ivory Gate.	Hubert Parry.
O, ye Dead. ⎱ The Song of the Ghost. ⎰	Old Irish.
Le Départ de l'Âme.	Old Breton.

The more obvious idea of illustrating ghostliness in the human voice is by giving an inhuman, unearthly sound to the ghost-voice, and this is right where the ghostly presence is sinister (as in the "Erlkönig" or "Waldesgespräch"), but there are many instances, such as the Parry song above, where such ghost-voices bring a message of tenderness or consolation to the mortal who is left on earth; or where, as in "O, ye Dead," the ghost speaks with a passionate longing to return to earth and the friends of life. In such songs the ghost must sound human, the man unearthly — that is, in colour. In a song like Schumann's "Auf das Trinkglas eines verstorbenen Freundes" the whole atmosphere must be saturated with the dead man's presence, though the dead man never appears or speaks.

(10) *Songs of Question and Answer*.

These, though nominally in this form, are generally deeply introspective and serious. Stanford's "Broken Song" has been spoken of earlier (p. 82).

Harold Darke's "Uphill," a setting of Christina Rossetti's famous poem "Does the road wind uphill all the way?" and the old Breton "Le Départ de l'Âme" are variants in question-and-answer form of the conversation between the Body and the Soul or the Mortal and Immortal. In these also the first is frightened and "unearthly," the second comforting and human, in colour. Compare also Hubert Parry's "Proud Maisie." The ordinary serious *Question and Answer* song is rather rare. Folk-song is full of its lighter forms such as "Spinnerliedchen" (in which the question is implied) and all the variants of "Where are you going to, my pretty maid?" etc.

(11) *Humorous and Quasi-humorous.*

Humorous songs are, providentially, part of the singer's accepted répertoire, but woe betide him if he tries to be funny! The daub of a brush or a dig in the ribs will turn a bit of humour into a caricature, and laughter to resentment. The music is painter and caricaturist in one, and wants no help from outside. Hammering in the joke with a "Sold again!" and a slap on the back is not appreciated any more on the concert platform than in the smoking-room of a club. The number of actually comic songs in legitimate music is small. On rare occasions an audience actually bursts out laughing, and then — it is to be hoped — at the words and music, not at the performance. The delightful exaggerations of the old English "Crocodile" or the Hibernianisms of the "Soliloquy" are funny enough in all conscience by themselves. To drive the point home by further exaggeration of delivery is akin to bidding the audience laugh at the mouth of a blunderbuss.

Dialect is a great asset in humorous songs, and here the English singer (or composer or poet) has no shame. The stereotyped three-verse English poem (?) about Pat and Molly and the Pig is turned out by the hundred yearly in this country, set in $\frac{6}{8}$ jig-time by the English composer, and sung with roast-beef accent, unblushing effrontery and oleaginous self-satisfaction by the English singer.

The following are excellent examples of humorous song :

The Laird of Cockpen.	Hubert Parry.
The Bells of Clermont Town. ⎫ Mary. ⎬ A. M. Goodhart. The Sailor's Consolation. ⎭	
The Crow. ⎫ Daddy-Long-Legs. ⎬ ("Cushendall.") C. V. Stanford. The Old Navy. ⎭	

and a large number of folk-songs. The Gilbert and Sullivan operas are, of course, full of them.

Under this heading also comes that class of song, half-humorous, half-affectionate, such as :

Johneen (an Irish Idyll). ⎱
Did you ever? ("Cushendall.") ⎰ C. V. Stanford.

The little song-cycle referred to — Stanford's "Irish Idyll" — contains in its six numbers an example each of the *Contemplative — Ode, Atmospheric, Reminiscence, Humorous — Affectionate, Question and Answer*, and *Dramatic — Contemplative* groups. In the last song half the scene is played in *Question and Answer* form in Canada, the other half in the "silver waters of the Foyle" at home.

SONG–CYCLES

Song–cycles at first blush would seem to demand a long chapter to themselves. To treat them in detail, which is the only way to discuss them adequately, is impossible in this particular volume for reasons of space. Song-cycles, as a matter of fact, are but the song in large. As the song in large is permeated and governed by an atmosphere of its own to which all the detailed phrases are subordinate, and to which at the same time every detailed phrase contributes, so on the higher scale, *mutatis mutandis*, the song-cycle takes the place of the song and the song of the phrase.

Song-cycles, for some reason, are almost invariably tragic. The singer of the "Dichterliebe" must be standing at attention before the first note is played, with his mind saturated with the tragedy of what is to come. He need not be afraid; the happiness of the opening will not be marred by the tragedy of his mood. His contrasts will be all the better for it. He must remember that if the audience does not know the course of that unhappy love-story, he does. He may act it, and paint every moment of every mood of every song with all the vividness he knows; he may shudder at the "dreadful hollow," call Maud from her dancing to the "woodbine spices" and "musk of the rose" in the garden, or dash his head against the stones in "Dead, Long Dead"; yet he knows the tale of the "Dichterliebe" or of "Maud" by heart before ever he begins to sing it. To him, at least, there are no secrets there: the atmosphere of the *whole* wraps him round.

Song-cycles are essentially the *Song of Characterisation* in large. The singer must assume the character for the time being, and act the story and its moods dramatically as it moves along. But no moment, no single song, must for purpose of passing effect be allowed to step out of the picture of the *whole*. Overelaboration and cheap effect ruin the song-cycle as inevitably as the song! Song-cycles are mostly written in the historic or dramatic present, and, as said above, must be acted. But however faithfully the acting be done, it is impossible to feel them as other than *reminiscent*. The music of the song gives a sacredness to the great human tragedies which the mind shrinks from violating. We see them on the stage, and are profoundly moved by them; but ennoble them with music and we instinctively uncover our heads. We think of them as having been, and keep them from the limelight.

THE SINGING OF FOLK-SONGS

Folk-songs for the purpose of this book must be held to cover *traditional* tunes generally, whether they be genuine folk-songs, ballads, or old tunes to which modern words have been put.

They are one and all governed by one rule of paramount importance.

Main Rule I. Never stop the march of a song.

This is the key to folk-song singing.

The folk-song, or traditional song, is generally very beautiful in form, and so dependent upon balance in structure — with long curving melodic phrases — that the slightest break tilts the balance over and brings the whole structure to the ground. If you break the phrases in the middle to take breath — and even in places where in the ordinary song it would be legitimate — you break the lines of the song whose structural design depends upon unbroken continuity.

The modern art-song is written and phrased by the composer to fit the words; but with the traditional song this is not necessarily so. Some have come down to us, words and music alike, as they stand; but a vast number (especially the greatest of all — the Irish) as tunes alone, and to these modern words have been put with masterly dexterity by such men as Thomas Moore of old and Alfred Perceval Graves of our own time.

In many cases the poet — especially in the case of fiddle-tunes — has had to dispense with the name-idea of the tune and write an entirely original poem. Thus, at the hands of one or other of the afore-mentioned:

"If the sea were ink" becomes "Lay his sword by his side."

"O what shall I do with this silly old man?" becomes "Colonel Carty."

"She hung her petticoat up to dry" becomes "One at a time."

"Whish the cat from under the table" becomes "Katey Neale."

"The twisting of the rope" becomes "How dear to me the hour."

"Leather-bags Donnell" becomes "The Alarm."

"Better let 'em alone" becomes (appropriately) "The Kilkenny cats."

Many, such as "The little red Fox" which was originally a quick dance tune and altered by Moore to the splendid "Let Erin remember," have been completely metamorphosised by the modern words and have consequently no textual characteristics to guide the singer. Processes have been reversed; words have been fitted to existing music, not music to words. Therefore *by the music alone* the singer must be influenced. He may use all his reserves of tone-colour and word-illustration to give point to the text, but the beginning and end of all things in the folk-song is *the integrity of the musical phrase.* Let the singer take the Stanford setting of the lovely old Irish tune "My love's an arbutus" (A. P. Graves), and experiment for himself.

My love's an ar - bu - tus By the bor-ders of

Lene, So slender and shapely In her gir-dle of green, And I

mea-sure the plea-sure of her eye's sap-phire sheen By the

blue skies that spar-kle through that soft branching screen.

Let him first sing it breathing only at X, and then
breathing both at XX and X, and compare the two.
The first will live; but no matter how dexterous the
breathing, and infinitesimal the time taken to do it,
the second will be dead.

Simple as it appears, and simple as it should sound,
no branch of singing is so difficult as *accompanied* folk-
song. Woe betide the singer whose lungs fail or whose
rhythm halts. In a modern song it can be forgotten;
in a folk-song never. The integrity of the phrase,
however long, is its very essence, and before its march,
words and prosody values and all else go down like
ninepins. If, as in a fiddle-tune like "Quick! we
have but a second," there is no time to breathe in
the course of the tune, the whole tune must be sung

in one, *without breathing,* or not sung at all. There
can be little doubt that many of the unaccountable
changes of time in folk-songs — the interpolation of a
$\frac{9}{8}$ bar into a $\frac{6}{8}$ song, or a $\frac{5}{4}$ bar into a $\frac{4}{4}$ song or
vice versâ — were due to some physical difficulty in
phrasing on the part of some exponent, past or present, most properly noted down by the present-day
collector. They have either put in an extra beat
when singing *unaccompanied* — as the true folk-song
singer does — to give themselves time to breathe or
to emphasise a certain syllable; or have left out a
beat that was not syllabically provided for, thereby
in either case altering the structure in a way that at
once becomes remarkable when the song is given an
accompaniment; as, for instance, in "Barbara Ellen":

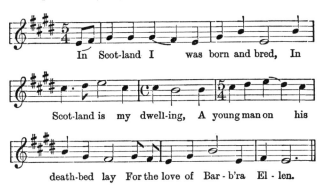

from Cecil Sharp's "Folk-songs from Somerset" (written
in $\frac{5}{4}$ time), where in the fourth bar the singer could
evidently not be bothered with three beats on the
second syllable of the word "dwelling" and simply
left out one, thereby automatically converting the bar
into an interpolated $\frac{4}{4}$.

It is the *addition of the instrumental accompaniment* which makes Main Rule I. imperative to the modern singing of folk-songs, for, strange to say, to *unaccompanied* folk-song singing it does not apply.

Unaccompanied folk-song singing is one of the most remarkable things in music. It breaks every rule of the art. It is the most *ad libitum* type of performance it is possible to imagine. The true singer of the people is born not made. He will drop an obvious beat here and put it in there. He will hold back a phrase to superimpose an ornament (this applies far more to Ireland, where ornament is used with an Oriental freedom, rather than to England where it is rare). He will dwell upon unaccented notes in unaccountable places. He may so juggle with the time that it may be impossible to give it a time-signature. The tune itself may be so peculiar that it may demand no arbitrary key-signature! Witness the old North of Ireland tune "My Lagan Love" (arranged by H. Hamilton Harty, poem by Seosamh MacCathmhaoil) :

love-sick len - an-shee She   hath my   heart   in thrall;

Nor life   I owe, nor   lib - er - ty,   For   Love   is lord of   all.

(To the eye this looks almost like an improvisation.
By the individual native singer it would probably be
further ornamented *ad libitum*. In the setting it is
treated hypothetically as C major, because it happens
to end on C. What its time-signature is, goodness
knows!) The singer may, in short, give himself a free
hand, break every rule and just sing; and yet he has a
rhythm of his own so strong that it sets the heart of
the trained singer leaping, so subtle that it defies
imitation — wholly fascinating, wholly unlearnable. It
is Nature as opposed to Art. No man who has not
got it in his blood, and who has not lived with it in
his youth, can ever acquire it. The further he travels
along the road of his art, the further he leaves that
astounding sense behind.

Unaccompanied folk-song singing at its best is only
given to a few to hear. It does not, unfortunately,
enter into the practical side of the public singer's life,
and its rules, or absence of rules, are not for him. He
has to do with the *accompanied* folk-song where the
great rules of rhythm apply most rigorously of all, and
not to one song but to all. Certain songs, such as
dramatic ballads ("The Twa Sisters o' Binnorie") or
ghost songs ("O, ye Dead") may be lent by folk-song
to modern art just to show how well they knew long

ago all there is to be said to-day; but folk-songs they will always be and very proud of it.

The *accompanied* folk-song is the hardest thing to sing in music — the hardest, and, from the point of view of study, the best to the man who is master of his technique. It depends upon the golden rule that except for exceptional reasons — the close of a song or some peculiar contrast — *you must never break the phrase in a folk-song, however long.*

If you cannot achieve that, like the Kilkenny cats — "Better let 'em alone."

The same advice may be respectfully offered to the young composer. To him, judging by recent experiences, the folk-song merely represents a ready-made vehicle for meretricious harmonies and polyphonic colour-illustration. The setting of folk-songs is so difficult that it may be almost legally assigned to the old hand. The master of the art does not flaunt his technique in your face or bejewel his Madonnas. He knows that here "beauty when unadorned is adorned the most," and that the secrets of folk-song setting are enhancement of rhythm and *economy* of material.

PART V

THE MAKING OF PROGRAMMES

(FOR reasons of space only Recital programmes will be discussed here. The greater includes the less.)

In the making of programmes, the singer starts with one advantage. He and his audience have the same object in view — to give or receive the *maximum* amount of interest with the *minimum* amount of fatigue, mental and physical. How can this be done?

By constant change.

The singer has only one voice. That voice may be many-coloured and of various degrees of power; it may be inspired by a vivid imagination and backed by a strong will, yet it remains one and the same voice. The change must be supplemented in the programme.

In the making of recital programmes, the singer must keep in mind the following ten essentials:

(1) Variety of Language.

(2) Change of Composer (except in the case of a group).

(3) Chronological Order.

(4) Change of Key.

(5) Change of Time.

(6) Classification of the Song.

(7) Style of Technique.

(8) Change of *Tempo* or Pace.

(9) *Crescendo* and *Diminuendo* of Emotion.

(10) Atmosphere and Mood.

(1) Change of *Language* is an undoubted relief to both parties. Apart from this, no translation, however good, can fulfil the conditions which inspired the composer. In the case of Hungarian songs and the like, where the language has dictated the rhythmic accentuation of the music, the language must be looked upon as Hungarian even though the words sung are English.

(2) In the miscellaneous programme change of *Composer* is advisable for change's sake. This does not apply to groups of songs, deliberately made, or to song-cycles.

(3) The value of *Chronological Order* is sentimental rather than real. It shows *en miniature* the progress of the Song. It should be considered here in conjunction with No. 9 (the Emotional Line). Songs have unquestionably progressed emotionally, and chronological order fits them into their places in the scheme of things. It bears somewhat the same relationship to Song in general as the *Opus* numbers to the individual composer, and helps to record development.

(4) Change of *Key* is of supreme importance. There is nothing so tedious, so "filling," as a succession of songs in the same key. If by inadvertence the singer places three consecutive songs in the same key in his programme, let him get his accompanist to transpose the middle one. The loss in tone through transposition will be far less detrimental to effect than the monotony of key.

A minor, however, can always be followed by the

major in the same key. It feels, in fact, the right thing. In the programme (p. 230) it will be seen that the "Doppelgänger" is followed by "Dort in den Weiden." The former ends in G major, it is true, but the song throughout is so much in the minor and so steeped in tragic gloom that one thinks of it as in G minor.

(5) Change of *Time* is also very important. It is difficult to ensure, for there are not many times to ring the changes with, and so many songs are written in $\frac{4}{4}$ or $\frac{3}{4}$ time that the field is narrowed. Happy is the man with a *répertoire!* Time changes are, however, not so all-essential to a programme as the changes of key.

(6) The *Classification* of Songs was treated in the last chapter. It will be noted that, with the "overlapping" songs, the singer has plenty of classes to choose from.

(7) Styles of *Technique* have also been discussed earlier (p. 145). All songs that are *sung* are *bel canto* in technique; *declamation* applies mostly to *dramatic* effects; *diction* to songs requiring *purity* and *agility* of pronunciation; the latter are naturally never deeply moving, their highest emotional sense, as a rule, being *happiness*.

(8) Changes of *Tempo* are not only legitimate but essential. A sudden contrast of pace may *startle* an audience (in itself an excellent thing), but it will not hurt like violent contrasts of sentiment. This can be explained better by

(9) The *Emotional Line* (so-called for want of a better word). The singer must not keep his hearers or himself too long on the stretch. They cannot stand it mentally, nor he mentally and physically. It is not

Q

the actual amount of voice used, or noise made, which counts for fatigue; it is the want of relief from mental strain. The three Harper's songs of Schubert, when sung consecutively, are (from their uniformity of sentiment) far more tiring to both singer and audience than half a recital programme properly ordered.

The principle of the *Emotional Line* is the arranging of the component songs in a series of *emotional crescendos* and *decrescendos*.

In the first part of the programme (p. 230) it will be seen that the top of the emotional *crescendos* comes upon

> Der Doppelgänger,
> Auf das Trinkglas,
> Silent Noon,

and in a lesser degree on

> Tap o' th' Hill.

"Mary," by Goodhart, a delightful little song in itself, would, if placed after the "Doppelgänger," set every nerve ajar. No singer with any faculty of discrimination could perpetrate such a sequence on the excuse of contrast. "Dort in den Weiden," on the other hand, gives both singer and audience exactly the right tonic for the nerves shattered by the "Doppelgänger." A simple little *optimistic* tune, the very antithesis of the other, it comes out like the sun after a month of London. Then Wolf's "Der Rattenfänger," though it makes a great noise, far greater than the deeply intimate "Auf das Trinkglas," which precedes it, has no emotional interest whatever, and provides a positive *decrescendo* of sentiment.

The line then rises to "Silent Noon." "Mary" again

could not come after "Silent Noon" without a jar. It could not come after "Tap o' th' Hill," the deep feeling of whose close attunes it to "Silent Noon," were it not that the greater part of "Tap o' th' Hill" is just a happy quasi-humorous little story.

In "Corinna's going a-Maying" there is no great emotion. It is, as it should be, all youth and high spirits. Its object in the programme in that particular place is to leave both singer and audience in a good temper if they have got thus far without coming to blows.

(10) Finally, *Atmosphere*. This has been fully treated already. It is closely associated with Main Rule II. It entails the singing of the song *as a whole*, in accordance with its classification.

In the case of song-cycles like the "Dichterliebe" (16 songs) the singer has half his programme ready-made and stereotyped for him. Here the audience makes up its mind to the worst, and, with the singer, accepts it as a whole and looks for no change except in detail. After such a first part, the second part of a programme *cannot possibly be too light*.

(It might be mentioned, in parenthesis, that it is well to provide for late-comers with a group of two or three songs before beginning the serious work of the song-cycle. One late-comer can destroy the mood both of singer and audience, and wreck not merely the individual song but the entire cycle. The singer would also be well-advised to ensure beforehand against the perambulations of programme boys; and the audience would pay the highest compliment to song and singer alike if it refrained from applause till the cycle *as a whole* is finished.)

Short song-cycles, such as Stanford's "Irish Idyll" (6 songs), should come at the *end* of Part I., and their style should be borne in mind in the making of the preceding part of the programme. The above little cycle, though full of variety, is permeated with an atmosphere of *deep affection* either for the country or the individual. The preceding songs should, therefore, ᴇ phasise the *declamatory, sinister, technical, rhythmical* or some other alien style, so that the pure emotion of the song-cycle when it arrives may come with a breath of fresh air and have that particular field all to itself.

In all programmes fatigue is the great enemy; therefore they should err on the side of lightness. This applies particularly to Part II., in which there should be a gradual *diminuendo* of mental strain. For this purpose folk-songs and traditional songs are admirable. After a long first part made up of classical and art songs, ancient and modern, grave and gay, German, French, Italian or English, declamatory or *bel canto*, with, maybe, a song-cycle or two thrown in, something is required in the second part that will allow the singer to relax his strain, and the audience its intellectual vigilance. Folk-song fulfils the conditions. Simple, straightforward, essentially rhythmical — melody pure and simple, with direct emotional appeal — it comes like the pipe after dinner at the end of a long day's work. There is no relaxation in them to the singer vocally — for they are the hardest things in music to sing well — but they act like magic on the tired brain; and to the public singer, possessed, as he invariably is, of physique, that is all that counts. With folk-songs the same conditions of programme-making apply as in Part I., but in a slightly less degree emotionally.

It is not suggested that the second part of a recital programme should be arbitrarily devoted to folk-songs; merely that folk-songs, of the many alternatives, "fill the bill" particularly well. The only real essential of such second parts is that they should be light.

To the superficial observer all this may seem to smack of our old enemies over-elaboration and self-consciousness; but it is not so. The old hand makes such programmes by instinct! Any singer possessed of the means and the necessary experience could turn out half a dozen of them in the course of a morning. The fact also remains that for the want of such careful manipulation the ordinary programme does not "come off."

In programme-making, the programme must be made first, and *made naturally*, and manipulated afterwards to conform with conditions. If the names of the songs in the programme overleaf were lost, it would perhaps be possible with infinite laboriousness to construct another programme to fit the analysis, but the result would be a triumph of artificiality. The programme must be made first without regard to any but the main conditions and "licked into shape" later.

[The programme given here was written down (as a specimen short recital programme) exactly as it stands, except for two slight alterations in Part I. and one in Part II. It was a thoroughly *natural* programme, and worked out into the analysis of itself. The alterations were in no way vital; they were made solely to conform with the conditions specified in this chapter.]

It must be obvious that the making of interesting

PROGRAMME

SONG.	LANGUAGE.	COMPOSER.	DATE.	KEY.
"Les petits Oiseaux" . . .	French	Traditional	16th cent.	E maj.
"Già il Sole dal Gange" . .	Italian	Aless. Scarlatti	1659–1725	F maj.
"Der Doppelgänger"	German	Schubert	1797–1828	G min. -maj.
"Dort in den Weiden" . . .	German	Brahms	1833–1897	G maj.
"Auf das Trinklas eines ver- storbenen Freundes"	German	Schumann	1810–1856	D maj.
"Der Rattenfänger"	German	Hugo Wolf	1860–1903	F♯ min.
"Silent Noon"	English	R. V. Williams	Present	D maj.
"Tap o' th' Hill"	English	H. W. Davies	Present	D♭ maj.
"Mary"	English	A. M. Goodhart	Present	C maj.
"Corinna's going a-Maying"	English	Ernest Walker	Present	D maj.

		Arranged by	TRADITIONAL
"Der Kukkuk"	German	C. V. Stanford	F maj.
"Shepherd, see thy Horse's foaming Mane"	Hungarian	F. Korbay	C maj.
"Will you float in my Boat?"	Irish	C. V. Stanford	D♭ maj.
"Remember the Poor" . . .	Irish	C. V. Stanford	G maj.
"I know where I'm goin'" . .	Irish	Herbert Hughes	G♭ maj.
"A Ballynure Ballad" . . .	Irish	Herbert Hughes	C min.
"Twankydillo"	English	Lucy Broadwood	G♭ maj.

FEBRUARY 24TH, 1911.

TIME.	CLASSIFICATION.	STYLE OF TECHNIQUE.	PACE.	EMOTIONAL[1] LINE.	ATMOSPHERE.
$\frac{4}{4}$	Contemplative (Ode)	Bel Canto	Andante		Devotional
$\frac{3}{4}$	Rhythmical	Bel Canto	Allegro		Very bright
$\frac{3}{4}$	Reminiscence (Dramatic and Atmospheric)	Declamatory (Bel Canto —Diction)	Very slow		Intensely tragic
$\frac{2}{4}$	Contemplative (Narrative)	Bel Canto (Diction)	Lively		Happy
$\frac{4}{4}$	Ghost (Reminiscence Atmospheric)	Bel Canto	Rather slow		Deep feeling
$\frac{6}{8}$	Characterisation	Bel Canto	Very lively		Very bright, almost with swagger
$\frac{3}{4}$	Reminiscence (Atmospheric)	Bel Canto	Andantino		Deeply emotional
$\frac{2}{4}$	Narrative (Contemplative)	Bel Canto (Diction)	Allegretto Vivace		Bright, affectionate and emotional
$\frac{5}{4}$ $\frac{7}{4}$	Humorous	Diction	Andante con grazia		Chaff
$\frac{2}{4}$	Ode (Rhythmical)	Diction	Allegro leggiero	‖	Youth and happiness

AIRS

TIME.	CLASSIFICATION.	STYLE OF TECHNIQUE.	PACE.	EMOTIONAL LINE.	ATMOSPHERE.
$\frac{2}{4}$	Humorous	Diction	Allegretto		Nursery Rhyme
$\frac{2}{4}$	Dramatic	Declamatory (Diction)	Allegretto		Barbaric and concentrated
$\frac{3}{4}$	Characterisation	Bel Canto	Allegretto grazioso		Emotional and happy
$\frac{4}{4}$	Devotional (Contemplative)	Bel Canto	Andante		Deeply Religious
$\frac{2}{4}$	Contemplative	Bel Canto	Moderato	◇	Reverie
$\frac{5}{4}$ $\frac{4}{4}$	Rhythmical	Diction	Allegro giojoso	‖	Humorous
$\frac{3}{4}$	Characterisation	Bel Canto	Con spirito	‖	Jolly

[1] The intensity of the various emotions is expressed by the actual size of the *crescendo* mark. The mark ‖ signifies absence of such emotional intensity.

programmes, and the fulfilment of their conditions, demand not only versatility, but, above all things, a *répertoire*. A dozen "popular" songs about angels, or organs, or Oriental swamps, or Pat and the Pig, will not carry the singer far, while the stock florid aria moves the listener to yawns as automatically as the latest new story at his club. A *répertoire* is not got without hard work, and hard work is the breath of life. Good programmes mean research and study, and intimate acquaintance with the masters of music. They are the true friends of interpretation, for they open the doors to individuality. No man can learn the whole of a great *répertoire* by imitation from anyone.

There is no intention in this chapter to demand arbitrarily the fulfilment of the above rules. They are but the detailed statement of a "counsel of perfection" which experience has shown it to be useful to aim at.

PART VI

HOW TO STUDY A SONG

To the old hand the studying of a song is mechanical; he probably does not know how he does it, and can only arrive at it by a process of analysis. It is, however, possible to simplify it for the new-comer.

There are various rules to be recommended, which are given below in their proper order, but they all belong to two Main Rules, representing respectively the *end* and the *means — interpretation* and *technique*.

Singing is the driving in double harness of the musical phrase and the literary sentence, and the pair must not only show off their paces but must *keep within the posts*. There is a natural antagonism between the two which endangers those posts at every step, and if the driver has not learned his business he will come to grief. We must presume that for the end — *interpretation* — he has at his disposal the means — *technique*.

Main Rule I. (the end).

Classify your song; in other words find out what it is all about.

Main Rule II. (the means).

Find your fundamental rhythm and absorb it.

Of these the second, though placed there, is in reality the most important. It has been urged throughout this book that the music's the thing, and that the rhythm is the thing in the music. The aesthetic meaning of a song, to which Rule I. refers, and its interpretation are dependent upon that rhythm and own it as their master; yet, though they have no separate existence without it, it is only a means to their ends, and must come second in study.

The song must be classified first, because the treatment of the rhythm must be reconciled to its atmosphere and consequent mood. If, for instance, the song be classed as *rhythmical*, the rhythm takes sole charge, making use of any help it can get from outside in the way of tone-colour and occasional word-illustration; whereas, if the song be classed as, say, *atmospheric*, the mood will dominate the situation, and the rhythm with its *rubato* satellites will come in as friendly, though indispensable, supporters of the scheme — a purely family affair in which the pater-familias supplies the funds and joins in the fun without directing the proceedings.

To recommend the singer to study his text first (Rule I.) would seem superfluous, if the call of the music were not so seductive. He rushes in and rattles off purely musical effects which will not bear juxta-position to the text; then in his disappointment he attributes the faults of his own "slap-dashness" to the shortcomings of the music. He has without knowing it made the song without words into a form of absolute music, with the name to give it a programme. The composer may be trusted to be true to his text, but that will not primarily classify the song for the singer. Most songs classify themselves, but if he is in any doubt the *master-phrase* (p. 17) will soon come and show him the way.

It is not enough to realise the rhythm; it must be *absorbed* (Rule II.). It is not enough for the singer to feel it as a slow $\frac{4}{4}$, or quick $\frac{6}{8}$, or anything else; it must so saturate his senses that all his effects and illustrations come under its spell and keep within its limits; otherwise the song becomes a series of detached observations ornamented by the addition of music. It is not a question of avoidance of difficulties, but a deliberate policy.

SUBORDINATE RULES

1. *Learn the song in rough.*

The singer need not trouble himself about effects at first. They will, of course, keep rushing into his mind as he learns. They are generally *musical* effects, and instinctive, and, if they are sound, will not be forgotten; they will turn up again at the right time.

2. *Memorise it.*

There will be no senses to spare for effect, or lilt, or magnetism, or illustration or anything else, if the eye is on the printed page. Let him memorise the song *in the rough* first, and then proceed to polish it up.

3. *Polish it musically first.*

The beginner who imagines that interpretation, when he gets to it, will be all dramatic illustration of his text, is doomed to disappointment. Half, if not more, of his effects will be purely *musical.* This applies not merely to the *bel canto* class, but to all songs.

Let him take the memorised phrases in detail and see what each is capable of *musically* in the way of *rubato, crescendo, diminuendo,* etc. He must not be led away by any artificial translation of the text into the music (cf. Consistency, p. 182). The composer has probably done that best, if he knows his business, and consistent phrasing is more important to him and the song than petty illustration.

4. *Reconcile the phrasing to the text.*

This is where the fun begins. In the great song there is no reconciliation to be made. They are joined together for better or for worse for all time, and the contemplation of that union is one of the principal assets of musical education. It is his own phrasing which the singer has to reconcile to the words. The music carries its effects inherent in its lilt; life begins for the singer when he begins to apply the results of his musical research for his own illustrative purposes. This is where individuality comes in. He can now tell a story, paint a picture, adumbrate an atmosphere, portray a character, act a drama, or recall a scene, in the words of poetry ennobled by music — *but one and all must be within the limits]* *(between the posts) of the rhythm of the song.* That is why the *musical* phrase has to be studied first. If

the fundamental rhythm will not permit of an illustration *musically,* no vocal sleight-of-hand can juggle
it into verisimilitude. The singer has means in plenty
at his disposal—tone-colour, word-illustration, *crescendo*
and *diminuendo, rubato,* pauses, climax and anti-climax,
and all the other helps to drama. Let him play the
great game fairly and *sing the song as a whole.*

A song is a little thing; it is over in a moment,
and there is no time to atone for early faults. Stop
it long enough anywhere and for any purpose, and the
attention will wander from the song to the singer; when
that happens, the song is dead. Rules 3 and 4 must
determine for him both his *style or styles of technique*
and *points of climax* or *anti-climax.*

5. *Absorb the accompaniment of the song.*

This is part of Main Rule II. (p. 92). The
accompaniment is not merely the business of the man
at the piano; it is the singer's as well. To sing the
song as a whole will be impossible if one-half is left out.
Every effect should be agreed on and rehearsed with
the accompanist; and together, share and share alike,
singer and accompanist must contribute to the whole.

To sum up,

> Find the Atmosphere of the Song;
> Sing the Song as a whole;
> Sing it as you speak it.

Let us now apply these rules to individual examples.
For reasons of space they must be limited in number,
and for reasons of copyright they must, with one
exception, be chosen from the classics. Two of them
are contained in the specimen programme (p. 230).
Ex uno disce omnes.

DER DOPPELGÄNGER. F. Schubert

(In this and the following songs many expression marks have been inserted for which the composer is in no way responsible. These are simply guides to method, *post hoc* applications of the result of analysis, used to illustrate the author's individual reading. The student should on no account adopt any one of them as his own unless it fits in with his conception of the song as a whole. Any other advice would be contrary to the very creed of this book, which has preached individuality as the great essential. This does not apply to word-illustration or the resumption of *tempo primo*, which are part of interpretation in general and should be followed in every song.)

"Der Doppelgänger" is a triumph of economy. It is but a couple of pages of music in length; it consists of a few simple phrases accompanied by simple chords. Yet in those few bars there is a dramatic *crescendo* of tragedy which has never been approached in music, and which fairly entitles it to be considered the greatest song in the world.

Let us take our rules in turn and apply them.

Main Rule I. *Classify your song.*

Read it through and visualise the scene.

The man is standing in front of the house where his sweetheart used to live. He watches, as though hypnotised, the window where he saw her night after night in the old days. The window is dark to-night; perhaps the house is empty; who knows? The agony of remembrance holds him in a vice.

She has gone long ago, but there is the house, *just*

as of old, and there stands a man staring up at it, *just as of old,* wringing his hands in misery, *just as of old.*

The moon comes out; he catches sight of the man's face; he cries out in horror — the face is his own! He breaks down, and with bitter taunts curses his own pitiful image for coming to mock him with the vision of all he suffered on that self-same spot in bygone days. Then, under the burden of remembrance — of the *tempo felice* and *miseria* — at the words *in alter Zeit* his head drops into his hands, and he collapses like a house of cards.

We hardly need to look for master-phrases here; but here they are:

> In diesem Hause wohnte mein Schatz.
> Here is the house where she used to dwell.

and

> So manche Nacht in alter Zeit.
> (literally) Night after night in those old times.

Where could it belong but to *Reminiscence?* Yet it is saturated with the atmosphere of tragedy, and is hardly less *Atmospheric.* Like many other songs of reminiscence it is in the present tense, but it is the *dramatic* present, not the historic — tragedy staged and played by the singer as he sings.

What a chance for him! *Reminiscence, atmosphere* and *drama,* all in one, concentrated into two pages of music!

MAIN RULE II.   *Find your fundamental rhythm.*

The rhythm of the "Doppelgänger" is felt not seen; there is no way of physically impressing it upon the senses. The simple chords are vertical and give a single down-beat. But that rhythm is like a tiger.

For the first twenty-four bars it crouches motionless; at the words *da steht auch ein Mensch* it begins to move stealthily forward; twice, at the words *vor Schmerzensgewalt* and *meine eig'ne Gestalt*, it seems about to spring, but it is not till *so manche Nacht* that the end comes. From the first forward gliding movement to the final leap the rhythm presses forward gradually but inexorably. Both singer and accompanist must have absorbed it into their very being. The dramatic effect of the whole depends upon the subconscious strictness of that rhythm in the first twenty-four bars and its subsequent *gradual* working to the climax. Let the words *da steht auch ein Mensch* be started a shade too fast or too slow; there will either be no pace left in reserve for the climax, or the effort to move the song along will have brought the physical climax too soon.

SUBORDINATE RULES

(Rules 1 and 2 apply to all songs alike.)

3. *Polish it musically first.*

The purely musical effects in this song are practically *nil*. It is essentially a song of interpretation, and the musical effects are so closely wedded to the aesthetic and dramatic as to be one with them. For this reason we will take this rule and the next together.

4. *Reconcile the phrasing to the text.*

The phrasing of the first twenty-four bars belongs to the "quiet pools" (*vide* p. 43). Whatever currents of rhythm there are, are below the surface; the surface is glassy smooth. The dead-level of semi-consciousness must dominate those bars; through them all the man *stares* fixedly. There is no room for *rubato* here. Once, in the eleventh bar, a diminutive *tenuto*, or push, sung *pp*, on *mein* suggests a throb of his heart, and this is followed, in the twenty-first bar, by a slightly longer similar push on *demselben*, which confirms the first musically (cf. p. 74) and links up the present and the past psychologically; but otherwise not a sign of incident or emotion.

From *da steht auch ein Mensch* music and drama push on gradually with a feeling of impatience to see the man's face, through the horror of recognition, *crescendo* to the climax on *so manche Nacht*, where the pent-up waters of misery break the dam to pieces, and memory— *in alter Zeit* — floats down, drowned, upon the tide.

The *colour* and *style of technique* also go hand in hand. The first twenty-four bars belong to that style referred to (p. 148) as the peculiar combination of *bel canto* and *diction*, "where in the anaesthesia of the tragedy all consciousness of the outside world seems to be lost." The words are half-spoken, half-sung, with dead-alive monotony. At *da steht auch ein Mensch* the technique turns to *diction* pure and simple — the man has woken to consciousness; the action begins — they are almost hissed out under the breath. This turns immediately — *und ringt die Hände* — to *declamation*, the *colour* expressive of intense suffering. Then the same process again — *diction* to *declamation* from *mir graust es* to *Gestalt*, the *colour* beginning with a shudder and ending with a cry. Then from *du Doppelgänger* a concentrated fury of pain (*declamation*, the *colour* sounding physically concentrated), leading to the physical climax — *so manche*

R

Nacht — where the barriers are swept away. Here the singer must sing the F♯ *ff*, as loud as possible consistent with beauty of sound; he must take special care that the two previous *fortes* on *Schmerzensgewalt* and *Gestalt* respectively are not overdone. He must leave a reserve for the final *manche Nacht;* otherwise he will have anticipated his climax.

Then comes a distinct *Kunst-pause.* Then a complete change of everything, technique to pure *bel canto*, colour to pure pathos — the singer must be sorry for the actor's lonely figure—the rhythm to practically *ad libitum* with a portamento down from the E to the D (to show the sinking of his head into his hands) with a general *diminuendo*, a sense of futility, a fading-away to nothingness to the end of the last note of the final symphony.

The singer has many opportunities for word-illustration. These are indicated by italics.

Sing the song as a whole.

5. If ever the singer should sing mentally through his rests it is here. Long before that opening "doomed" chord is played he should be ready watching the window, and through every sung note and every played chord of the song up to the climax he should *stare.*

The "Doppelgänger" should be rehearsed by singer and accompanist till both know the reading by heart. The accompanist should above all take care that the opening chords are played cleanly; even a suspicion of spreading will destroy the hidden rhythm. They should sound "doomed" with the tone of a deep bell, not of a harp. He too must regulate his *crescendos* and *fortes* so as to leave a reserve for his climax.

The song is practically *recitative* throughout.

DER DOPPELGÄNGER

HEINE.   F. SCHUBERT.

Still ist die Nacht,

es ru-hen die Gas-sen,   in die - sem Hau-se

wohn - te *mein* Schatz;

sie hat schon *längst* die *Stadt* ver - las - sen,

doch *steht* noch das Haus auf dem-sel - ben Pla*tz*.

Still is the night; the streets are silent;
    Here is the house where she used to dwell;
'Tis long since my darling left the town,
    But the house still stands, that I knew so well.

Who is the man there standing and staring,
    And wringing his hands in the torture of woe?
I tremble, as slowly he turns towards me —
    'Tis my own pale face the moonbeams show.

Thou ghostly double! wan, sorrowful comrade! —
    Why dost thou mock my sighs and tears,
The throes of love so often suffer'd
    On this dear spot in bygone years?

                                    PAUL ENGLAND.

ER DER HERRLICHSTE VON ALLEN.   R. SCHUMANN
        (Frauenliebe und Leben. No. 2.)

This song is entirely *musical* in its effects; it is in
this respect the exact opposite to the "Doppelgänger."
The composer has used them with masterly cunning
to express the meaning of the text, but none the less
the interpretation is governed by purely musical effects
which might apply to any other equally emotional
words.

(I.) It is a frankly ecstatic panegyric in the form
of a contemplation; an unblushing *Schwärmerei* or

hero-worship, touching in the pure unselfishness of its very *abandon*. It belongs therefore to the *Contemplative* and *Ode* groups. The poems of the little song-cycle from which it is taken give us an impression of a girl, very young, very emotional and very ecstatic in her outward expression; we can with justification picture her here walking up and down her room, clasping and unclasping her hands, pressing them to her heart, and every now and again in a moment of rapturous enthusiasm throwing her arms wide to the hero of her dreams.

(II.) The rhythm is self-evident; it is, in fact, itself the song's interpretation. The effects throughout are effects of rhythm, accent, *rubato* and *tempo primo* and nothing else.

(3) and (4). This being so, Rule 3 is going to take charge and Rule 4 has got to make the best of it. Every effect of illustrative tone-colour, style of technique, word-illustration, etc., must be subordinated to the lilt of the musical phrase. The strength of the rhythm and its lilt lies in its series of *rubatos* and subsequent *tempo primos*. Many of these *rubatos* have been written by the composer in his usual form (by a *ritardando* with no instructions as to the return of the *tempo primo*); but the most powerful of all are the turns with the implied push or *tenuto* on the note immediately succeeding them. There are fast and slow turns in music of which the latter are almost somnolent in their peacefulness (*vide* Brahms's "Feldeinsamkeit"); but the quick turn (which is the form used in this song) is one of the most potent forward-driving forces we know of. Its effect is one of stimulation, and, when combined with the *tenuto*

on the succeeding note and the consequent holding-up
of the rhythm, it can only be described as intoxi-
cating. Such turns are sung in strict time, or even
quicker than strict time (so as to anticipate the beat),
giving an impression as though the singer *dashed*
at her accents. The subsequent *tenuti*, or pauses,
conform strictly in their treatment to those mentioned
earlier (pp. 74 and 160) the second being always
stronger in power, and longer in duration, than the first.
(This will be shown overleaf.) It may be safely said
that the interpretation of the song depends upon the
treatment of the turns.

The style of technique is *bel canto;* it is pure singing
throughout. The *colour* is ecstatic, with little waves
of humble-mindedness running through it, such humil-
ity, with its colour, disappearing the moment the
singer turns from introspection to speak of her hero.
The *word-illustration* is practically done by strength of
accentuation. The initial consonants of the words
falling on the first accents must ring like steel; where
such words begin with a vowel the note must be struck
in the very centre and on the instant of the down-beat,
while the vowel must be purity itself. *Er* must be
monosyllabic with a vengeance (not *ay-ur*), or the whole
point of the song, the concentration on the worship of
the hero, will be lost. The open *a* (ah) of *klares* must
be the true open *ah*. *Klaw-res* or *klah-ures* would
be suggestive of the prize-ring or the Bier-kneipe.
Wandle, wandle deine Bahnen is sung with a suspicion
of tears in the voice. The *w* can, by pure pronunciation, carry a subtle emotional effect of the sort.
Likewise the alliterative *h* in *Er an meinem Himmel,
hell und herrlich, hehr und fern* gives a splendid

chance to the word-painter. The strong accents throughout on words like *Er, Auge, Tiefe, Stern, Herrlichkeit,* etc., etc., and the pure singing of such vowels as in *Er, Auge, wandle, Bahnen, traurig, Stern, Wahl, hohe,* etc., are illustrations in themselves. The metre is trochaic, and many of the trochees are written on two even crotchets ♩ ♩ ♩ ♩ They
Lip-pen, Au - ge, etc.
must, of course, be given common-sense values, not equal pressure values, and sung as true trochees.

There is no definite *note* of climax; the climax is the return to the original subject *Er der Herrlichste.* The leading up thereto is in the accompaniment. The interpretation of the song *as a whole* is again evenly divided between singer and accompanist. There are pitfalls at every step for both of them. Over-elaboration and cheap effect have bird-limed or poisoned every other bar to destroy the *little* interpreter. The composer has written three *ritardandos* to begin with. He has also written or implied ten turns; each of these carries a pause or *tenuto,* holding back the forward movement for an appreciable time. Not one of these must be held a fraction of a second too long, and not one must be taken out of the picture. As the singer sings each turn she may clasp her hands or clutch at her heart or throw her arms out to the beloved image, but here the *music's the thing* and the rhythm inexorable; therefore through every *ritardando* and every turn the singer must be conscious of the *push-on,* of the drive, of the song as a whole. The accompanist is here all-powerful. He cannot, it is true, pull the singer down from her *tenuto* perch after the turns, but he can and *must* pick up the *tempo primo* when he gets it to himself, and swing it along with all the *abandon* of which he is capable. (The addition of the seventh to the chord in the main places of resumption carries the idea of forward swing inherent in it.) He must remember too that the effect of the final climax is in his hands. The words *brich, O Herz, was liegt daran?* have been sung sadly, almost *piano.* He has three bars in which to rise from a *piano* to a *forte,* three bars in which to bring the singer back from introspection to ecstasy, and resolve the song as a whole. When he gets to his big octave C let him give it with a will; its rhythmical effect will be far more important to the interpretation than anything the voice can do.

Finally, if the spreading of the arms on the last turn (*fester*) be prolonged with extra *abandon,* as is quite legitimate, the subsequent *ritardando* must be very slight.

ER DER HERRLICHSTE VON ALLEN
(Frauenliebe und Leben.)
CHAMISSO.

R. SHUMANN.

Muth.

So wie dort in blauer

a tempo.

p

Tie - fe   hell und herr - lich je - ner

colla voce.

Stern,   al - so Er   an meinem

a tempo.

Wand - le, wandle dei-ne Bah-nen, nur be-

trach - ten dei - nen Schein, nur in

De - muth ihn be - trach - ten,

se - lig nur und trau - rig sein.

Hö - re nicht mein stil-les Be - ten, dei-nem

Glü - cke nur ge - weiht; darfst mich

8

Wür - digste von Al - len darf be -

glü-cken dei - ne Wahl, und ich

will die Ho - he seg - nen vie - le

tau - send - mal: will mich

freu - en dann und wei - nen,

se - lig, se-lig bin ich dann, soll-te

Er,   der *Herr*lichste von *Al* - len,   wie so

mil - de, wie   so *gut*!   *Hol*-de

Lip - pen,   kla-res *Au* - ge,   hel-ler

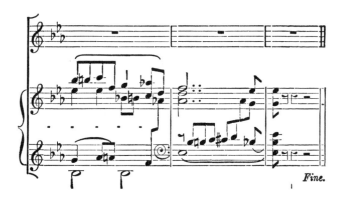

Fine.

He's the noblest of the noble ! None so gracious, none so kind !
Courtly greetings, fearless glances, lofty soul, and constant
    mind !

See, aloft in yonder heaven, clear and radiant shines a star ;
So for me my hero shineth, calm and radiant, clear though far.

Lonely, lonely as I watch thee on thy glorious journey go,
Still my humble heart shall bless thee, happy in my secret woe.

Thou wilt never hear my pleading while for thee on heaven I call ;
Not for me, the lowly maiden, shines the fairest star of all.

Only one, the best, the noblest, wilt thou choose to share thy way,
And for her, the highly favoured, gladly, gladly will I pray.

Whether joy be mine, or sorrow, blessed still I count my lot,
Though this heart should break with anguish — Break, oh
    heart ! it matters not.

                            PAUL ENGLAND.

AUF DAS TRINKGLAS EINES VERSTORBENEN
FREUNDES R. Schumann.

They told me, Heraclitus, they told me you were dead.
They brought me bitter news to hear and bitter tears to shed.
I wept when I remembered how often you and I
Had tired the sun with talking and sent him down the sky.
WILLIAM CORY.

This is essentially a *Ghost* song; it is filled with the presence of the dead man.

That presence is *Atmospheric*, for the dead man does not appear.

It is steeped in *Reminiscence* — unlike the "Doppelgänger" — of happy days.

It is *Dramatic*, for the picture is vivid; but the drama is played in silence. The very *crescendos* are mental. The words are thought, not sung.

The whole scene is dominated by the waiting for, and coming of, the dead man.

Those who have read the poems of the late T. E. Brown will remember the famous *Epistola ad Dakyns*, in which the poet asks his old friend, if he survives him, to go back to the three places they both had loved best in life, and promises him, if it be allowed him, to return again and meet him there in spirit.

So it is here. The singer is here by appointment. He and his friend had agreed in life that if one died the other should come back to the old spot, where together they "had tired the sun with talking," at midnight before the anniversary of his death, and as the hour sounded take down the dead friend's glass, fill it and drink to his memory; and then, so surely as they had stood together in life, so surely they should stand together in death.

You see him come in and take down the glass. It is covered with cobwebs — it has stood there for a year. He fills it with the golden wine and waits. The starlight is all the light in the room. The clock begins to strike. He stands up and drinks. As he drinks the moon comes out from the clouds. At the last stroke he puts the glass back on the table — empty. He stands quite still and listens. Far away in the bowl of the empty glass he hears the echo of the bell. He knows his friend is with him, that the friendship of life has survived even death itself.

> No friendship dieth with death of any day.
> Wie nichts den Freund vom Freund kann trennen.

The song is written in a series of detached sentences, generally accompanied by a *ritardando* and pause at the end of each. These *ritardandos* give a sense of deliberateness — the scene is played thus at every anniversary. In the pauses the living man — the singer — gazes intently at the glass; the watcher — the listener — holds his breath. The rhythm is simple and unadorned. There is not a *rubato* in the whole of it; the dramatic pauses, as in all songs of detached phrases, take charge of the movement. There is not a real *forte* or big note anywhere in the song. It must be sung throughout with pure *bel canto*, and with a colour of "reverence" up to the words *in deinem theuren Blute!* In the sentences immediately preceding, and leading up to, these words, there must be a gradual *crescendo*, emblematic of the faith that is in him. So far he has mentally *spoken;* he has addressed an

invocation to the dead friend's glass. At this point
he drinks; the rest *happens*. Therefore after the
words *in deinem theuren Blute!* there must be a long
pause — he slowly raises the glass to his lips. Then
the accompaniment must start with a different colour
— very far away and mysterious (the atmosphere is
dominated now by the dead, not the living). That
colour must be assimilated in its turn by the voice;
it should sound remote, impersonal and very gentle.
The moon sails out slowly from the clouds; midnight
strikes. The strokes of the bell can be illustrated by
infinitesimal, perfectly even, pushes on the notes.
There should be a distinct *ritardando* in the bar
before the words *Leer steht das Glas*, as he slowly puts
the glass back empty; and a pause on *Leer* ("empty")
to show that he has carried out his share of the
compact. Then another pause in the accompaniment,
followed by a silence — he *listens*. Then *pp*, with all
the beauty of far-away colour the singer can command,
the echo must die away to the end. The voice is but
the echo of the bell, its message for him alone.

Here again the accompanist must play his chords
clean, not spread, throughout; but it is impossible to
tabulate this song according to dogmatic rules. The
whole is so atmospheric, so intimate, so dependent
upon mood, that singer and accompanist in their
regard for the greater will give small heed to the less.
For that very reason it has been chosen as an example.
The nobility of the subject demands reverential treatment; coarse vowels, exaggerated consonants or final
slurs would be sheer *lèse-majesté*.

AUF DAS TRINKGLAS EINES VER-
STORBENEN FREUNDES

J. KERNER.

R. SCHUMANN.

Ernst, ziemlich langsam (with deep feeling, rather slow).

Du herr - lich Glas, nun stehst du leer,

Glas, das er oft mit Lust ge - ho - ben;

die Spin - ne hat rings um dich her in -

- dess den düs - tern Flor ge - wo - ben.

ritard.

Jetzt sollst du mir ge - fül - let sein, mond -

p

hell mit Gold der deut-schen Re - ben!

In dei - ner Tie - fe heil' - gen Schein schau'

ich hin - ab mit frommem Be - ben.

Was ich er-schau' in dei - nem Grund, ist

nicht Ge-wöhn-li-chen zu nen - nen. Doch

wird mir klar zu die - ser Stund', wie

nichts den Freund vom Freund kann tren - nen. Auf

die - sen Glau-ben, Glas so hold ! trink' ich dich aus mit

ho - hem Mu - the. Klar spie - gelt sich der

Ster-ne Gold, Po - kal, in dei-nem theu-ren Blute !

Still geht der Mond das Thal ent -

Now drear and empty dost thou stand,
   Brave glass, so often drained by him,
The spider with ill-omened hand
   Weaves gloomy webs about thy rim.

Yet thou to-night once more shalt glow
   With moonlight gold of Rhenish wine,
While, gazing in thy depths below,
   My soul thy secret shall divine.

The wondrous things that there I see,
   I dare not to the world deliver, —
Yet now indeed 'tis plain to me
   That life nor death true friends can sever.

Thus, noble glass, I drain thee dry,
   Strong in the faith these thoughts impart;
The gold of all the stars on high
   Shines clear from out thy glowing heart.

The silent moon through heaven climbs, —
   Sternly the midnight hour is tolled, —
Void stands the glass, — but solemn chimes
   Still linger in its crystal hold.

PAUL ENGLAND.

THE CROW[1]   C. V. STANFORD.

(From the Song-cycle, "Cushendall.")

This is a humorous song in the form of a contemplative appreciation. It is a doffing of the hat to the successful rogue.

The master-phrase (if one be required) is "My faith, the bird is sly" — not the slyness of Uriah Heep; there is nothing humble about him. He is the Richard Hardie of *Hard Cash*, with a little of Pecksniff thrown in where diplomatically advisable.

The mood is given in the words:

> "He cocks his head wi' knowin' look,
> And scans ye wi' his eye."

You must sing the whole song with your head figuratively tilted N.E. or N.W., and with a cold glitter in your eye. There is not a *laugh* in the whole of it; it is a series of appreciatory *chuckles*. There is nothing consciously humorous about either of them; both are out for business.

The point of the song, psychologically, is, of course, the emphasising of the analogy; and musically the enforcement of that emphasis by the strict *a tempo* phrase. (This is indicated by italics throughout.) So vital is the return to this strict *tempo primo* that judicious *rubato* must be deliberately made throughout the song in order to emphasise the re-entry of the

[1] Published here by kind permission of the composer. The poem, from *Pat M'Carty his Rhymes*, is printed by kind permission of the author and Mr. Edward Arnold.

rhythm on the analogy. Not only so, but a short
pause on the note immediately preceding it — even
when on an unimportant word and beat, or even after
a *fermata* — should invariably be made in order to
"hold up" the rhythm and make it the more welcome
when it comes.

The style of technique is *Diction* throughout, and
there are numbers of points for word-illustration; but
these and all the other points in the song must be
subtle — "the bird is *sly.*"   There is not a real
"effect" in it anywhere. It depends upon its alternating elasticity and rigidity of rhythm, on the
pointed forefinger of its diction, and the suppressed
humour of its tone-colour. It belongs rather to the
"miniatures" of song, and is proportionately hard to
sing.

The opening symphony begins the illustration for
us delightfully. Its alternating *staccato* and *legato*
quavers puts us into the stride of the bird and the
pose of the man — a series of "stalks" and hops. (He
must be very careful where he puts his feet, and keep
a strict look-out for traps — he is a great stickler for
legal etiquette.) The accompanist should play them
as though he meant to hop across the bar-line, peered
over first, and then thought better of it. The success
of the song depends as much upon the humour of his
touch as on the colour in his colleague's voice.

"He wears a bla*ck* coat a' the week."

Here you may dwell upon the picture a very little —
just enough to suggest the sanctimoniousness of his
pose and the deliberation of his investigations. This
is done as much by diction and colour as by *rubato.*

The *ck* and *c* of "bla*ck c*oat" can be given a *caw! caw!*-like suggestion of sound.

> "A long at*torn*ey *beak*
> For *pok*in' into things."

You can get a world of sarcasm into the accentuation and colouring of the italicised syllables. They should suggest a squint, or ocular concentration associated with some form of ugliness. The *ea* of "beak" should sound a thin — almost nasal — *ee.*

> "He takes some interest in lands,
> And talks a kind o' jaw."

He rubs his chin here. Land is not really in his line at all, but if he can be of any service he will make it his business, etc., etc. He doesn't know how he happened to alight in that field. You can convey amused contempt into the *j* of "jaw," and by a pause on the word can suggest the length of the sermon.

> "He looks by or'nar stern and grim."

The crotchets, sung *f* (the *f* is only comparative; there are no real *fortes* in the song; the gentleman in question is a blackmailer not a highwayman), must be rigid in time-values, expressive of his rigid probity and Puritanism.

> "He's certain verra wise."

You chuckle to yourself as you say it. You are the only person in the secret; as in the "Leiermann" you watch the drama from afar.

The next few sentences up to the word "grab" are but explanatory amplifications of why you chuckled.

There is nothing to stop for here, so they are sung in strict time.

> "He cocks his head wi' knowin' look,
> And scans ye wi' his eye,
> As if to read ye like a book."

You will find that, if you have been in the mood throughout, your head will probably (quite unconsciously) have assumed the necessary tilt physically! The picture to those acquainted with the two prototypes is so vivid that the physical response is spontaneous.

> "My faith, the bird is sly."

You chuckle out loud here. "Did you threaten him with the law? my dear Sir, the law is his business!" Admirable things scarecrows as resting-places and vantage-points from which to spy out the land!

> "He gives ye help mayhap *some* days."

A certain grudging generosity of tone here; on the "some" an implied shrug of the shoulders — the days are few and far between, and only when it pays him to play your game.

> "And kills a slug or twa."

Here is a chance for sarcasm. Killing is the type of his benefactions, a slug or two the limit of his subscriptions to the commonwealth. He has no objection to paying Paul by robbing Peter.

> "But costs ye dear in other ways,
> Just like the man o' law."

This is the moral — the summing up — of the whole song. You can do what you like with it, provided

you do not take its *piú lento* out of the rhythmical
picture. The two last bars, with the *a tempo* crotchets
(instead of quavers) and the end in the *major*, are a
stroke of genius. You have shot at him and missed
him; tried to poison him — he has "cocked his head
wi' knowin' wink"; laid traps for him — he has
"scanned them wi' his eye." You can see him
returning thanks after dinner on his election as mayor
of the town. With his hand in his breast he modestly
disclaims all merit but patriotism. His comprehensive
rascality, ready for anything from malingering to
Spanish treasure, is known to every man in the hall.
Every one of them has tried to get the better of him,
but with the tricks of his trade — graft, blackmail,
intimidation, and the rest — he has beaten them all in
the end.

THE CROW

JOHN STEVENSON.   C. V. STANFORD.

If men have got their counterparts A-
mong the birds, the craw Wi' a' his cuteness and his arts Is
sure the man o' law. He's got the impudence and cheek That

talks a kind o' *jaw*   That no man liv - in' un-derstands, Just

a tempo.

like the man o' law.

He

a tempo.

looks by or' - nar stern and   *grim,*   He's cer-tain ver - ra

a tempo.

like the man o' law.

He

cocks his head wi' knowin' look, And scans ye wi' his eye,   As

a tempo.

if to read ye like a book,   My faith, the bird is *sly.*

APPENDIX

HOW TO BREATHE

MORE voices are ruined by breathing faddists than by any other type of vocal charlatan. The damage is all the more dangerous for being insidious. Bad voice-production tells its tale in time and gives the student physical warning; he sees the cracks in the walls and clears out. But the foundations may be rotten, and he will never know.

Singing is the finest natural exercise in the world. It may be hard to separate cause and effect, but the fact remains that public singers, in spite of late hours, bad air, nervous strain, and a life spent in trains and hotels, are, as a rule, unwarrantably and disgustingly healthy. Like all public men, they suffer under the scandal-monger with "private information"; their non-appearances are put to the credit of inebriety or some other vice, and the public lifts its hands in horror at the story — as silly· as it is patently untrue — and wonders how they do it. The English singer — with rare exceptions, who disappear young — is as sound a good fellow in mind as he is in body.

It is not the faddist who has made singing the finest exercise in the world; it is Nature. She has told all living things to breathe in the way that gives them least trouble. She did not bother fish or fowl, cat or dog, horse or man with minute instructions about uvulas, epiglottises, vocal cords, diaphragms or *Rectus abdominis*, to be carefully followed before barking, or neighing, or singing, and least of all before breathing. She told them to go ahead and begin; and she made no exception in the case of singers; she merely remarked that if her ordinary gifts were to be used for special purposes they should be specially developed. It is that special development which makes singing such a splendid exercise; for not only does it expand the lungs, but it abnormally strengthens the great supporting muscles of the body.

The faddist's victims can be told at a glance. One fills merely the top of his lungs and suffers negatively from shortness of breath. He can be recognised by his pumping shoulders, short phrasing, and generally harassed expression. The other forces his lungs downwards and outwards into a place where they have no call to be, and to the serious detriment of those organs which are there by right. The so-called "abdominal" breather can be marked by his flat chest, distended stomach, dyspeptic, pasty complexion, and relentless *tremolo*. To teach such breathing to men is bad enough; to women it is almost criminal.

Healthy breathing is simplicity itself. There are only two things to think of, and of these the first has nothing *directly* to do with breathing.

(1) *Lift the chest as high as ever it will go, and keep it there throughout the whole process of singing, from a single note to a song.*

The object of the raised chest is three-fold:

(a) to get it out of the way and give free play to the lungs;

(b) to increase resonance of the voice;

(c) to give "presence."

The concert singer should stand (with chest expanded and the weight thrown on the forward foot) as straight as a dart, not like a crescent moon or the comic tramp of the music halls. A flat-chested Wotan would be laughed off the stage; why not an "abdominal" Elijah? This raising of the chest is independent of breathing. It can be raised as easily when breathing out as in. When once it is up, it must be kept there till the song is done; this will be found to have a remarkable effect on Main Rule II. — *Sing mentally through your rests.*

(The beginner, or singer, who has not been accustomed to raising his chest, will probably suffer at first from pains below the lower ribs. This is all in the day's work and is due to the unaccustomed use of fresh muscles; it will disappear in time.)

(2) *Breathe in the centre.*

Long phrasing depends, as stated earlier, not on the large amount of breath taken in, but on the small amount given out. It is but common sense to say that the economic control of breath should be given to the part of the body which has the necessary muscles to assume it. There is exactly such a muscle in the centre of the body, ready and willing for the work, and capable of developing unlimited powers of expansion and gradual relaxation. It works in conjunction with other great muscles in the centre of the body. What its name is does not matter a straw; let us call it the "breathing muscle" and proceed to place it.

Run your hand down your breastbone till you come to the end. Where it stops, the ribs branch off on either side. In the triangle between those ribs — with the apex at the end of the breastbone — there is a muscle which in its potentialities for breath-control can only be described as magnificent. It is in exactly the centre of the body, and when developed not only gives control of exhaltation but, with its friendly colleagues, supports the most important internal organs.

Now raise the chest as high as it will go; then breathe in (through the nose, whenever there is time) slowly. As you breathe *in*, the breathing muscle should expand *out*. (The beginner, from sheer "cussedness," has a tendency to do the opposite and, as he breathes in, to contract the muscle. This is fatal. He should keep his fingers pressed against the breathing muscle and, as he breathes *in*, should feel and see the muscle push his fingers *out*.) When the lungs are full and the muscle expanded as far as it will go, hold all taut in *statu quo* for an appreciable time; then press gently with your fingers on the muscle, and breathe out as slowly as ever you can, holding the breath back and relaxing the muscle by the slowest possible degrees. As the breath goes *out*, the muscle contracts *in*.

The training of that muscle to regulate its *relaxation* at the singer's will is the foundation on which singing is built.[1]

The singer who follows out these instructions will be conscious of a sense of disappointment when breathing *in*. The amount taken in feels infinitesimal; he has no run for his money; he has nothing to show for it.

[1] These instructions have been submitted to the highest expert opinion and endorsed as sound, physiologically, in every particular.

So much the better. If there is nothing shown there will be nothing seen, and breathing should be neither seen nor heard. A visible breather is like a "roaring" horse; he may do his work, but he is distressing. It is quite right that he should have nothing to show for it. His chest, being raised, does not press on his lungs, and there being no chest to push out, he fills his lungs instantaneously and automatically without feeling it; not only so, but if the chest is raised and the breathing muscle expanded, rib-expansion—on which good teachers quite rightly insist — follows of itself automatically, without any special attention.

If he still doubts, let him sit down quietly on a chair, relax his muscles, *drop his chest*, and then breathe in. The process is exactly the same as when his chest was raised; but it feels now like moving mountains. The power of it is athletically exhilarating, but akin to weight-lifting instead of throwing the cricket ball. If he will but concentrate his mind on the one great centre muscle and *on that alone*, and train it to follow his will, breathing will give up refusing its fences, and will carry him over anything he puts it at.

The rules of breathing have thus been simplified down to a single one. There is not a fad in the whole of it; like most sound things its working is not seen, and it makes no fuss. It is founded on common sense, and is the stock-in-trade of every singer who knows his business. It is only restated here because any attempt to interpret without it would be a farce. It has one inestimable advantage — it applies in every particular to men and women alike.

THE CLERGY AND INTONING[1]

Some Practical Considerations

IN a recent lecture on "Interpretation in Song," I dwelt strongly on the golden rule that singing must be speech in song, to attain which the following essentials must be observed: (a) purity of diction, (b) identity of texture in quality of the singing and the speaking voice.

I have been asked by the Editor of the *Church Family Newspaper* to contribute an article on the subject embodying what I said in the lecture, in so far as it referred to intoning. I said that in this great question of diction the English clergyman, in the matter of intoning, was one of the worst offenders. There are many brilliant exceptions, but in the main the statement is true, though he is not nearly so deserving of blame as the singer.

The clergyman (for the purpose of this article this must be taken to mean the intoning clergyman) and the singer have much in common. Their immediate object is the same — to move their hearers as deeply as possible. Their means are the same — the spoken word in song. But is that word truly or purely spoken in song by either one or the other? Not once in a hundred times. For the singer

[1] This article appeared in the *Church Family Newspaper* of December 30th, 1910, and is republished here by kind permission of the Editor.

there is no excuse whatever. Purity of diction in song and
musical declamation are as much part of his business as a
knowledge of the Scriptures is that of the clergyman. The
singer's neglect is due to sheer slovenliness, to the presumption that his voice is enough by itself, and that he
need not trouble to make himself intelligible to his
audience. Not so with the clergyman. He simply *follows
the line of least resistance* — for safety's sake. He is not
primarily a singer. The singing voice is to him an
extra. In many cases he has studied hard to train that
voice — often of beautiful quality — and with it has produced remarkable results, to his infinite credit; but he
generally fails in one great essential. His diction is not
pure. His intoning is not speech in song. There is no
slovenliness about it; he simply follows the line of least
resistance.

He listens to his voice ringing down the long aisles, and
his one idea is to preserve the pitch, the quality of tone,
the poise and the straight line — the equivalent to the
straight line in phrasing of the singer.

Clinging as he does to this ideal, every big clear vowel
feels to him a highwayman, every consonant a footpad
lying in wait to bring him to destruction, to hurl him from
the path, the path that winds over mountain and flood,
through meadow and wood; so he takes the safe way, the
"semi-detached" road lined on either side with Chats-
worths, Blenheims and Ivy Banks, and walks along the
middle of it, safe but suburban.

The pure vowel he dare not tackle, the consonant is
best not meddled with, so he shades practically all vowels
alike, dark or light, pure or modified, closed or open, to a
monotonous compromise, a mixture of "ah," "aw," "ur,"
and "oo," an indiscriminate hybrid, wholly indescribable
and hopelessly tedious — the sound which converts the

manly English pronunciation of "God" into a compromise between "Guard" and "Gurd." (In this, as in the examples below, the "r" is simply put in as a guide to the eye. It is not suggested that it is sounded.) On the same principle,

"man" becomes "mahn" (with a suspicion of "murn" in it),
"holy"   "   "hawly,"
"church"   "   "charch,"
"peace"   "   "pairce,"
"think"   "   "thairnk,"
"the"   "   "tha,"
"move"   "   "murv" (or "möve," the German modified ö),
"of"   "   "arv" or "urv,"
"porch"   "   "parch,"
"day"   "   "dair,"
"death"   "   "durth,"
"cloud"   "   "clard," etc., etc.

It will be seen that these pronunciations are one and all simply variants of, or modifications towards, the same single open "ah," the ground-vowel of singing, Nature's primary vocal sound — in short, they follow the line of least resistance.

I have chosen the above words at random, simply as examples of different vowel-sounds. Let any reader of this article make up for himself a sentence containing these words and then intone it, following with his eye, meanwhile, the values I have put to them, and let him ask himself if generally they are unjust. Let him then intone, say, the first three lines of "Paradise Lost." This is the way it would sound intoned by the average clergyman, the fact being borne in mind that in those vowels written down as representing the "ah" sound there is generally a supplementary "ur" tone quite impossible to express on paper:

Arv mahn's farst desorbairdyurnce ahnd tha frurt
Arv thaht furbeddern tray hurs martall tairste
Brahrt durth entor tha warld ahnd ahl ahr war.

This to the eye is a monstrous caricature, but intoned faithfully with the monotonous drawl which seems to be the only safe form of delivery, it will strike home. The same form of phonetic spelling applied to the opening prayers of the Church service would be even more convincing.

Now, why should the clergyman feel compelled thus to abuse the beauties of his own language? He does not speak it so. There are cases within my own knowledge where the constant habit of "least resistance" intoning has superinduced the "least resistance" habit in the speaking voice, and has earned for the individual the undeserved reproach of priggishness — the speaking voice of the curate of the comic stage; but in most cases his voice in speech is pure and clear. Why should it be otherwise when that speech is put into song? Is not "God" a beautiful-sounding word in itself? If you call it "Guard" or "Gurd," its majesty drops from it like a rotten garment. What more beautiful word for mere sound than "holy"? If you call it "hauly," it might belong to a comic chanty in a nautical operetta. Or the word "peace"? That pure, deep, closed vowel carries peace in the very sound of it. If you call it "pairce," you give it a sound of contempt, a curl of the lip like a snarl.

"Of man's first disobedience." Why should "man" be "mahn"? In the extremes of registers in singing, the various vowels automatically shade themselves to darker or lighter colours, or, when a note has to be sustained *forte* for a considerable time, the vowel has often to be broadened to give full tone; but this does not apply to the vowel when intoned. The intoner sings on a comfortable note in the middle of his voice. There is not the slightest excuse for altering the colour or texture of the spoken vowel.

"Disobedience." The word is a little model in itself. Two closed vowels, one at either end, and two absolutely pure, strongly contrasted vowels in the middle. (Compare it with "desorbairdyurnce.") Let the intoner speak that one word over many times with absolute purity of vowel values. Let him then intone it on his own particular intoning note. Let him then take the whole line and do the same, and so on with each line in succession, as far through the poem as he cares to go. Let him then take, say, the General Confession and treat it the same way, and each prayer in turn, and as he moves along, gradually the mists will lift, the sun come out, and the sky turn to blue. I will promise him that, the first time he sings the service thus, his magnetic sense will tell him that his congregation has suddenly begun to stand at "attention." The old familiar words which they have automatically assimilated, yawned over, or even slept through, have put on new life. There are fresh beauties in every line. The Collect for the day is at last audible and intelligible to them. They are awake, alert, listening not to the beautiful voice, not to the unicoloured, monotonous delivery, not to the man, but to what he says. That is the true object of interpretation, whether of Church service or of song.

And the terrifying difficulties, what are they? What is the intoner afraid of? That the larger and more violent movements of the actual pronouncing organs of diction will throw the tone off its balance, spoil the pitch, disturb the poise, and divert the straight line? They will do nothing of the sort. Pronunciation or diction is done practically entirely by fine closely associated movements of the lips and tip of the tongue round the teeth. If diction takes place farther back in the mouth, it is wrong. If the intoner produces his voice properly so that it has what is technically called the "nasal ring" (which does not in the least imply a nasal sound), and will sing his vowels quite pure, and will concentrate his mind, meanwhile, on that forward point — the tip of the tongue, lips, and teeth — he will find that the consonants follow of themselves. They are not nearly such stern foes to him as to the singer.

Let him then practise as above a speech from Shakespeare, or the various portions of the Church service, as easily and monotonously as he pleases, taking breath long before he really needs it, so as to avoid strain, but singing every word, vowel and consonant alike, *exactly as he would speak it.* The difficulties will vanish, the beauties of his language will stand out, the very mastery of it will fill him with pride, and the memory of the line of least resistance with shame. As the result of a fairly long experience of all sorts and conditions of singing — opera, oratorio, concert and Church music — I confidently assert that the man who held his audience the closest has not been the possessor of the most beautiful voice, but the man who *spoke* to them in song.

I have written all this with considerable diffidence. It may be said that it is none of my business. But intoning is very much part of every singer's business. We use it to study with and practise with, and as an invaluable thera-peutic for disordered technique. The words of any musi-cal passage which is difficult to sing should always be intoned first on a single note. Both singing and intoning should be speech in song. It is for this reason I trust our experience may be of some service.

Finally, I should recommend all intoning clergymen to take a few lessons from some well-known teacher of singing, one whom he knows to be sound, and whose pupils are a

testimony to his powers of turning out healthy voices and
pure diction. Let him eschew the faddist and the purveyor
of short cuts to technique (there are none), the patent
"voice-producer" and the anatomical-jargon man. The
less the pupil, clergyman or singer knows of the anatomy
of his voice the better. He wants to be taught how to
intone or how to sing, not to think of his uvula or vocal
cords and become self-conscious over their involuntary
movements. In this connection I should, however, strongly
emphasise one thing. The man who does not naturally
and with absolute ease breathe through his nose can never
either sing or intone. His nose is far more important to
him than his throat, and on the free passage of air through
his nose depends his tone, his pitch, his poise, and his
straight line. Ninety-nine cases out of a hundred of
"clergyman's sore throat" are due to the blocking of that
passage, and the consequent throwing of the sound into the
back of the throat, with all that that entails — loss of voice,
physical fatigue and mental despondency. If the clergyman is conscious of any stoppage or thickness there, let him
go to any one of the well-known specialists and have it
removed. It is generally a small matter surgically, but of
supreme importance vocally.[1] Half a dozen singing lessons
will do the rest. Then, with his diction pure and his intoning speech in song, his life will begin afresh. The
voice that was his enemy will suddenly be his friend; the
voice that was beautiful before will take on new colours;
the big congregation that he could not hold will suddenly
thrill his magnetic senses with its sympathy. Its every
member will at last hear him, understand him, and respond
to him.

[1] This applies even more strongly to singers.

INDEX

Abendlied (Schumann), 157; (Schulz), 183, 207.
Abschied, 145, 208.
Absent, yet present, 123.
Accent, 40.
Accompaniment, 46, 47, 86, 93, 99, 196, 242, 251, 267, 278.
Address, songs of, 201, 206.
Ah! Golgotha! 151.
Alarm, the, 217.
Alberti bass, 87.
Allmacht, die, 206.
Allnächtlich im Traume, 180.
An die Leyer, 149, 206.
An die Musik, 56, 206.
An meinem Herzen, an meiner Brust, 209.
An Schwager Kronos, 209.
Anakreon's Grab, 145.
And he journeyed with companions, 155.
Angels ever bright and fair, 61, 97, 102, 138.
Annie Laurie, 138.
Anti-climaxes, 179, 192, 237.
Appendix, 289.
Aria, the, 146.
Ariel, 61.
Ariosos, 151.
Arnold, Mr. Edward, 277.
Arpeggio, the, 47.
Art-song, the, 56, 80, 121, 199.
Art thou poor? 205.
Articulation, 119.
Artificiality, 16, 19, 182.
Aspirates, 113.
Atmosphere, 4, 13, 32, 36, 98, 200, 227, 237.

Atmospheric songs, 201, 202, 234, 239, 265.
Auf das Trinkglas eines verstorbenen Freundes, 211, 226, 230, 265.
Auf dem Kirchhofe, 14, 18, 202.
Auf dem Wasser zu singen, 87, 205.
Auf Flügeln des Gesanges, 48, 102, 207.
Ave Maria, 207.
Axioms, 195.

Baal recitatives, 154.
Bach, xi, 57, 58, 62, 83, 133, 145, 149, 151, 155, 172, 181, 193, 207, 209.
Balance, 46, 56, 71, 73.
Ballad, the traditional, 21.
Ballad, the accompanied strophic, 26, 218.
Ballynure ballad, a, 230.
Barbara Ellen, 219.
Baritone, the, 108.
Barnby, Sir J., 106, 114.
Bass, the, 107.
Beethoven, 156.
Beglückte Heerde, 84, 209.
Beiden Grenadiere, Die, 95, 203.
Bel Canto, 57, 60, 102, 121, 132, 145, 154, 166, 171, 200, 207.
Belle Dame sans Merci, La, 27, 202.
Bells of Clermont Town, the, 147, 213.
Bid me discourse, 207.
Birds in the High Hall Garden, 48.
Bishop, 207.
Blest Pair of Sirens, 159, 192.

301

Blow, Dr., 207.
Boat Song, 209.
Bois épais, 169.
Brahms, 14, 45, 52, 60, 72, 136, 152, 159, 194, 202, 205, 206, 208, 230.
Breathing, 6, 62, 65, 85, 142, 289.
Broadwood, Miss Lucy, 230.
Broken Song, the, 82, 162.
Brown, Rev. T. E., 265.
Brünnhilde, 107.
Bülow, Hans von, 10.
But the Lord is mindful of his own, 138.

Cadenza, the, 172.
Carissimi, 114.
Caro mio ben, 60, 145, 177, 207.
Carrying-over, 114, 169, 192.
Cathedral traditions, 107, 135.
Chamisso, 252.
Characterisation, songs of, 203.
Cheap effect, 32, 44, 77, 169, 193.
Chopin, 88, 167.
Chords, 54.
Choruses, 133.
Chronological order, 224.
Church Family Newspaper, 294.
Church service, 297.
Church's one foundation, the, 135.
Classification of songs, the, 198, 225, 233.
Clay, F., 161.
Clergy and intoning, the, 294.
Clergyman's sore throat, 300.
Climaxes, 179, 237, 251.
Colla voce, 53.
Colonel Carty, 217.
Come, blessed Cross, 57.
Comin' thro' the rye, 138.
Composer, the, 3, 35, 131, 175, 222, 224.
Consistency, 182.
Consonants, 7, 106, 113, 184, 295.
Contemplative songs, 201, 205, 249.
Continuity, 92.
Contralto, the, 45, 52, 60, 63, 106, 119, 176, 190.
Conventions, 35.

Corinna's going a maying, 85, 147, 208, 227, 230.
Cornelius, 20, 87, 100.
Corrymeela, 205.
Cory, William, 265.
Crocodile, the, 212.
Crow, the, 147, 213, 277.
Cushendall, 147, 205, 277.

Dactyls, 123, 189.
Daddy-Long-Legs, 147, 213.
Dakyns, Epistola ad, 265.
Danza, La, 61.
Darke, Harold, 181, 212.
Davies, H. Walford, 180, 204, 205, 230.
Dead, long dead, 165, 209, 215.
Debussy, 202.
Declamation, 146, 154.
Dekker, 205.
Départ de l'Ame, le, 211.
Detail, 71.
Diagnosis, 148.
Dialect, 213.
Dichterliebe, 2, 78, 100, 103, 122, 137, 146, 156, 203, 214, 227.
Diction, 146.
Did you ever? 147, 213.
Diphthongs, 111.
Dithyrambe, 208.
Doppelgänger, der, 122, 146, 148, 156, 204, 225, 230, 238, 248, 265.
Dort in den Weiden, 147, 225, 230.
Drake's Drum, 51.
Dramatic Songs, 202, 239, 265.
Du bist die Ruh, 136, 166.
Du bist wie eine Blume, 56, 179, 206
Du Hirte Israel, 83.
Du Ring an meinem Finger, 171.

Echoes, 162.
Ecoute d'Jeannette, 147.
Ein schön' Tageweis, 168, 210.
Ein Ton, 20, 87, 100, 102.
Elasticity of phrasing, 77, 167.
Elgar, Sir E., 160.
Elijah, 112, 138, 154.
Emotional line, the, 225.
England, Paul, 125, 248, 264, 276.
English singer, the, 35, 138, 143, 289.

Equipment, 4.
Er, der herrlichste von Allen, 248.
Erlkönig, der, 2, 14, 20, 42, 45, 78, 87, 95, 146, 147, 177, 179, 195, 202, 211.
Es blinkt der Thau, 204.
Ethiopia saluting the Colours, 15, 100, 165, 202.
Evangelist, the, 154.
Expression-marks, 193.

Facial expression, 16.
Faddists, 6, 289, 300.
Fairy Lough, the, 2, 67, 165, 202, 209.
Far and high the cranes give cry, 66, 69, 140.
Feldeinsamkeit, 12, 14, 18, 20, 42, 72, 136, 202, 204, 249.
Fermata, the, 74, 81, 82, 158, 168, 278.
Finish of a song, the, 71, 81, 176.
Flannel Jacket, the, 40.
Floodes of Tears, 207.
Florid songs, 57, 59, 132, 166, 200, 207.
Flowers that bloom in the spring, the, 78, 161.
Folk-songs, 199, 216, 228.
Folk-songs from Somerset, 219.
Follow a Shadow, 147.
Ford, Walter, 200.
Forelle, die, 203.
Franz, 177.
Frauenliebe und Leben, 122, 147, 148, 156, 203, 252.

Gabriel, Virginia, 198.
Gentle Maiden, the, 178, 210.
German language, the, 120, 137.
Gesänge, des Harfners, 205.
Ghost Songs, 211, 265.
Già il sole dal Gange, 61, 63, 208, 230.
Gilbert, Sir W. S., 78.
Giordani, 60, 145, 207.
Give, O give me back my Lord, 57.
Gluck, 157.

Goodhart, A. M., 147, 213, 226, 230.
Gounod, 31, 112.
Graves, Alfred Perceval, 132, 217.
Greise Kopf, der, 156.
Gretchen am Spinnrade, 202.
Gruppe aus dem Tartarus, 14, 208.

Haidenröslein, 147, 161.
Handel, 59, 60, 96, 145, 207.
Hans Sachs, 141.
Harty, H. Hamilton, 220.
Hatton, 123, 198, 206.
Have mercy, O Lord, 57.
Haynes, Battison, 205.
Heine, 243.
Henschel, Dr. Georg, 152.
Hey! Nonny No! 205.
How dear to me the hour, 217.
How does the wind blow? 209.
How to breathe, 289.
How to study a song, 233.
How willing my paternal love, 61.
Hughes, Herbert, 230.
Humorous and quasi-humorous songs, 212.
Hungarian, 140, 167, 224.
Hurdy-gurdy man, the, 98, 125.
Hymn-tunes, 133.

I hate the dreadful hollow, 156.
I know where I'm goin', 230.
Iambics, 117, 123, 188.
Ich grolle nicht, 178, 208.
Ich hab' im Traum geweinet, 156.
Ich kann's nicht fassen, 147.
Ich sende einen Gruss, 163.
Identity of Texture, 104, 142, 294.
If with all your hearts, 138.
Ihr Bild, 156.
Illustration, 51, 67, 99.
Im Rhein, im heiligen Strome, 177.
Imagination, 13, 48.
In der Fremde, 205.
Incidental high note, the, 138.
Individuality, 6.
Instrumental subject, the, 85, 99.
Interludes, instrumental, 88, 96.

Intoning, 142, 294.
Introduction, ix.
Irish Idyll, An, 67, 82, 203, 213, 228.
Italian language, the, 106.

Jesus, Saviour, 57.
Jewel Song, 91.
Job, 134, 192.
Johneen, 49, 213.
Junge Nonne, die, 161.

Katey Neale, 217.
Kerner, J., 268.
Key, change of, 224.
Kilkenny Cats, the, 217, 222.
Korbay, Francis, 15, 66, 102, 140, 146, 147, 162, 230.
Kukkuk, der, 147, 230.
Kunst-pause, die, 165, 178, 180, 192, 242.

Laird of Cockpen, the, 16, 176, 213.
Language, change of, 224.
Large, treatment of song in, 31, 64.
Lascia ch'io pianga, 60, 191.
Late-comer, the, 10, 227.
Lay his sword by his side, 217.
Leading-note, the, 179.
Legato, 48.
Leiermann, der, 2, 16, 17, 98, 102, 125, 148, 156, 202, 279.
Let Erin remember, 217.
Lied im Grünen, das, 208.
Lilt, 32, 38, 249.
Line of least resistance, the, 109, 119, 295.
Litanei, 170, 207.
Little red Fox, the, 217.
Little song, the, 77.
Lorelei, die, 211.
Lost Chord, the, 115.
Love is a Bable, 205.
Lover's Garland, A, 131, 175.
Lully, 170.

MacCathmhaoil, Seosamh, 220.
Madam, will you walk ? 25.
Magnetism, 4, 8, 31, 36, 94, 164.

Maid of Athens, 31.
Maiden, the, 176.
Main Rule I., 37, 216.
Main Rule II., 92, 196, 237, 291.
Main Rule III., 104, 184.
Mainacht, die, 72, 137, 205.
Mannerisms, 33, 166.
March, the, of a song, 37, 96, 101, 158.
Marseillaise, the, 95.
Martini, 171, 205.
Mary, 147, 213, 226, 230.
Masterphrase, the, 17, 102, 235.
Maud, 48, 156, 165, 191, 203, 209, 214.
Meine Liebe ist grün, 45.
Meine Seele rühmt und preist, 58.
Melisma, the, 59, 172.
Melody, 42, 73.
Memorising, 12, 235.
Mendelssohn, 48, 60, 102, 112, 154, 155, 207.
Messiah, the, 59, 108, 111.
Metre, 56, 104, 120.
Mezzo-soprano, the, 108.
Mikado, the, 78, 80.
Milton, 105, 205.
Minnelieder, 210.
Miscellaneous concert, 89, 94.
Moffat, Dr. Alfred, 41.
Mohàc's Field, 15, 20, 102, 140, 146, 162.
Mood, 13, 32, 36, 51, 234.
Moore, Thomas, 217.
Morgengruss, 46, 161.
Motion, 42, 43, 45.
Moto perpetuo, 83.
Müller, Max, 118.
Musical comedy, 111.
Musical notation, 121, 123.
My Lagan Love, 220.
My love's an arbutus, 217.
My name is John Wellington Wells, 78.

Narrative songs, 203.
Nasal ring, 299.
Nazareth, 112.
Neugierige, der, 122, 147.
Nightfall in Winter, 202.

Now sleeps the crimson petal, 207.
Nun hast du mir den ersten Schmerz gethan, 148, 156.
Nussbaum, der, 48, 209.

O rest in the Lord, 60, 63.
O ruddier than the cherry, 138.
O, ye dead, 211, 221.
Obbligato, Organ, x, 35, 55, 95.
Ode Songs, 206, 249.
Old Navy, the, 213.
Old Superb, the, 208.
Ombra mai fu, 207.
One at a time, 217.
Over-dramatising, 169.
Over-elaboration, 18, 33, 44, 77, 128, 169, 193, 229.
Over-sentimentalising, 169.

Pace, 7, 182.
Paradise Lost, 296.
Parlato, 78.
Parry, Sir C. H. H., 16, 131, 134, 135, 147, 159, 175, 183, 202, 205, 206, 211, 212, 213.
Pat M'Carty his Rhymes, 277.
Patchiness, 26.
Pause, the, 44, 74, 77, 158.
Personal recollections of Johannes Brahms, 152.
Petits oiseaux, les, 230.
Petrie Collection, 40.
Phonetics, 118.
Phrasing, 7, 43, 57, 65, 92, 236, 240.
Physique, 5, 228.
Plaisir d'amour, 171, 205.
Poise, 62, 101, 140.
Polysyllables, 184, 188.
Popular song, the, x, 198.
Portamento, the, 176.
Pressure-values, 124, 251.
Prima donna, the, 91, 207.
Programmes, the making of, 223.
Prolongation of note-values, 77.
Prosody, 56, 104, 120.
Proud Maisie, 212.
Psalm, the 104th, 192.
Psalters, 135.
Purcell, 121, 211.

Purity of Diction, 104, 294.

Quartet in A minor (Beethoven), 156.
Question and Answer songs, 82, 162, 211.
Quick! we have but a second, 83, 147, 218.
Quilter, Roger, 207.
Quintuple time, 167.

R, the letter, 115.
Rallentando, the, 168, 176.
Rattenfänger, Der, 88, 203, 208, 226, 230.
Recitative, 83, 146, 148, 242.
Rehearsal, 48, 74, 90, 237, 242.
Rejoice greatly, 59, 111, 145.
Remember the poor, 230.
Reminiscence, Songs of, 204, 239, 265.
Répertoire, 5, 232.
Rests, 74, 77, 103, 163.
Rhythm, x, 32, 38, 45, 56, 73, 121, 144, 149, 167, 182, 234, 239, 249, 266.
Rhythmical Songs, 57, 61, 132, 166, 200, 207, 234.
Rising phrase, the, 135.
Roadside Fire, the, 73, 160, 206.
Romantic school, the, 122.
Rondeau, the, 171.
Rose, die Lilie, die, 78, 146, 195.
Rossetti, Christina, 212.
Rossini, 61, 106.
Rubato, 77, 80, 158, 167, 169, 171, 234, 277.
Rubinstein, 204.
Rules, 37.

Sabbath morn at sea, 160.
Sailor's Consolation, the, 213.
St. John Passion, 172, 181.
St. Matthew Passion, 57, 149, 151, 154, 155, 172.
St. Paul, 155.
Samson Agonistes, 205.
Sands of Dee, the, 161.
Santley, Sir Charles, 31.
Sapphische Ode, 52, 60, 87.

x

Sarasate, 41.
Scarlatti, Aless., 61, 63, 208, 230.
Schmied, der, 208.
Schöne Müllerin, die, 46, 203.
Schubert, 2, 15, 16, 17, 20, 45, 46, 49, 56, 75, 76, 80, 98, 122, 125, 136, 145, 146, 147, 149, 156, 161, 162, 164, 166, 170, 174, 178, 193, 195, 202, 203, 204, 205, 206, 207, 208, 209, 226, 230, 238.
Schulz, 183, 207.
Schumann, 2, 16, 21, 45, 48, 56, 76, 78, 81, 95, 122, 146, 147, 157, 163, 171, 178, 179, 202, 203, 205, 206, 208, 209, 211, 230, 248, 265.
Selby, Luard, 80, 202.
Self-banished, the, 207.
Self-consciousness, 35, 44, 77, 101, 229.
Shakespeare, 105, 120, 299.
Sharp, Cecil, 219.
She came to the village church, 191.
Shepherd, see thy horse's foaming mane, 140, 146, 147, 230.
Silences, 164.
Silent Noon, 2, 14, 53, 227, 230.
$\frac{6}{8}$ time, 209.
Slur, the, 107, 176.
Soliloquy, the, 212.
Somervell, Arthur, 21, 48, 156, 165, 191, 203, 209.
Song-cycle, the, 92, 103, 214.
Song of the Ghost, the, 211.
Song of the Sou'-Wester, 209.
Soprano, the, 59, 108.
Sorcerer, the, 78.
Sostenuto, 60.
Spinnerliedchen, 147, 212.
Spondees, 123, 187.
Stadt, die, 164.
Staleness, 18.
Stanford, Sir C. V., 2, 18, 27, 49, 67, 82, 147, 162, 165, 170, 202, 203, 205, 208, 209, 211, 213, 217, 228, 230, 277.
Stevenson, John, 281.
Stockhausen, 195.
Straight line in phrasing, the, 42, 62, 101, 114, 140, 295.

Strophic songs, 81, 199.
Study, 1.
Style, x, 29, 36, 69, 95.
Styles of Technique, 145, 225, 237, 241, 250, 278.
Sullivan, Sir A., 78.
Syncopations, 52.

Tap o' th' Hill, 227, 230.
Tears, idle Tears, 204.
Technique, x, 4, 36, 57, 233.
Temperament, 5, 95.
Tempo, change of, 225.
Tempo primo, 80, 159, 161, 169, 171, 176, 251, 277.
Tenor, the, xi, 56, 58, 108.
Tenuto, 74, 158, 168.
Tessitura, 102.
Texture, 62, 104, 142, 294.
Thomas Kirche, xi, 58.
Through the ivory gate, 211.
Time, change of, 225.
Time-signatures, 121, 196.
To Althea, 206.
To Anthea, 123, 206.
To Lucasta, 206.
Tod und das Mädchen, der, 76.
Todessehnen, 205.
Todessehnsucht, 145, 207.
Tone, 57.
Tone-colour, 4, 19, 36, 48, 51, 83, 241, 250.
Tradition, 30, 194.
Traditional songs, 216.
Translations, 139, 224.
Tremolo, 290.
Trillo di natura, ⎫
      ”   agilità, ⎬ 19.
Trochees, 117, 123, 186, 251.
Trottin' to the Fair, 147, 170, 208.
Turn, the, 249.
Twa Sisters o' Binnorie, the, 21, 161, 202, 221.
Twankydillo, 230.
'Twas in the cool of eventide, 57, 151.

Una voce poco fa, 91.
Ungeduld, 15, 45, 174, 208.
Uphill, 181, 212.

Vagabond, the, 203, 208.
Vergebliches Ständchen, 45, 147, 194, 202.
Victorian period, 87, 106.
Vittoria! Vittoria! 114.
Voice-producer, the, 300.
Voice-production, 7, 142.
Von edler Art, 210.
Vowels, 7, 106, 184, 250, 295.

Wächterlied, 210.
Wagner, 113, 157.
Waldesgesprach, 16, 21, 146, 147, 202, 211.
Walker, Ernest, 85, 147, 205, 208, 230.
Wanderer, der, 75, 156, 162, 203.
Wandern, das, 49, 122, 208.
Wenn ich in deine Augen seh', 156.
Wesley, S. S., 133, 135.
When Childher plays, 180, 205.
When lovers meet again, 183.
When the flowing tide comes in, 106.

Where the bee sucks, 61.
White, Maude Valérie, Miss, 123.
Who'll come fight in the snow? 41.
Whole, treatment of song as a, 13, 26, 44, 237, 242.
Widmung, 45, 178, 206.
Widow Bird, A, 80, 202.
Wie bist du, meine Königin, 159, 206.
Will you float in my boat? 230.
Williams, R. Vaughan, 2, 53, 73, 160, 203, 204, 206, 208, 230.
Wirthshaus, das, 156, 166.
Wohin? 208.
Wolf, Hugo, 80, 145, 203, 208, 230.
Wood, Charles, 15, 100, 165, 202.
Word-illustration, 184, 242, 250.

Ye twice ten hundred deities, 211.

Zapateado, 41.